ROMAN BUILDING TECHNIQUES

ROMAN
BUILDING
TECHNIQUES

Tony Rook

AMBERLEY

IN MEMORY OF MERLE ROOK
1935–2012

Front cover: The House of Diana, a balconied tenement of Ostia, where commonly the brick facing of the *opus testaceum* was displayed.

Frontispiece: Roman soldiers shown building using rectangular blocks (possibly mud bricks) on Trajan's column.

First published 2013

Amberley Publishing
The Hill, Stroud
Gloucestershire, GL5 4EP

www.amberley-books.com

Copyright © Tony Rook 2013

The right of Tony Rook to be identified as the Author
of this work has been asserted in accordance with the
Copyrights, Designs and Patents Act 1988.

All rights reserved. No part of this book may be reprinted
or reproduced or utilised in any form or by any electronic,
mechanical or other means, now known or hereafter invented,
including photocopying and recording, or in any information
storage or retrieval system, without the permission in writing
from the Publishers.

British Library Cataloguing in Publication Data.
A catalogue record for this book is available from the British Library.

ISBN 978 1 4456 0149 6

Typeset in 10pt on 12pt Sabon.
Typesetting and Origination by Amberley Publishing.
Printed in the UK.

CONTENTS

	Introduction	7
1	Vitruvius	11
2	Building Types	14
3	Surveying	30
4	Foundations	33
5	Wood	36
6	Stone	40
7	Mud	46
8	Lime	50
9	Mortar & Concrete	53
10	Plaster Stucco & Fresco	56
11	Bricks	61
12	Masonry Walls	64
13	Roofs	70
14	Floors	77
15	Stairs	82
16	Windows	84
17	Doors	89
18	Metals	91

19	Water	95
20	Pipes	105
21	Hypocaust	109
22	Chimneys	114
23	Baths Come of Age	122
24	Thermae	126
25	Strength & Stability	129
26	Lintels	135
27	Arches & Vaults	141
28	Domes	147
	Acknowledgements	156
	Further Reading	157
	Index	158

INTRODUCTION

I tell this tale, which is strictly true,
Just by way of convincing you
How very little, since things were made,
Things have altered in the building trade.

Kipling, 'A Truthful Song'

Modern archaeology involves almost every science and, unfortunately, some of its practitioners seem willing and able to pontificate with confidence on a wide spectrum of subjects which are totally outside of their practical experience. Pointing to a few stones buried in the ground or a few dark patches in the soil, an academic will confidently tell us all about the building they represent. He will ask an artist or a computer expert to produce a highly convincing and structurally impossible picture of it. Even worse, he will get a few of his colleagues together who, for the entertainment of a television audience, will give a practical demonstration of their lack of knowledge, ability, skill and experience as builders.

The professional body representing archaeologists, the Institute for Archaeologists, includes a 'Buildings Special Interest Group', which acts, it says, 'as a forum for promoting the archaeological analysis, research, interpretation of *standing structures*'. The italics are mine. The specialist group makes no mention of bridging the gap between the sometimes flimsy, almost subjective evidence in the ground, and its interpretation as a drawing, a computer model or a real 'bricks-and-mortar' solid, upstanding, reconstruction. Building technology plays no great part in the training of archaeological students.

A gothic cathedral is a building. So is terminal five at Heathrow. One fundamental difference between them is that each is a product of its time. Their principal functions are very different, as are the materials from which they are made. The technology involved in construction has evolved and so has the psychology of the builders. An archaeologist who is excavating the remains of a building should know something about both the technology that is likely to have produced it and the ideas in the minds of the people of its time. Was a structure traditional, *avant garde*, 'retro', *à la mode*? Examples of conservatism persist, are recycled or implied throughout history. The Victorians built with gothic arches in cast iron. A Greek temple for Hera or a Roman one for Claudius imply the same 'post-and-lintel' construction, translated into stone, as their early prehistoric wooden prototypes. They are found on the imposing fronts of many town halls today, with pediments and columns, and as feeble imitations in Garden City architecture.

Kipling wrote the poem quoted above a century ago, and things *have* changed remarkably in recent times. I have often found myself wanting to write in this book 'nowadays' or 'in modern times' when, in fact, my memory is of things as they were when *I* was a lad. I remember, for example, when scaffolding was made of wooden poles lashed together (Fig. 32), when mortar was, very sensibly, made by mixing lime with sand, when a bricklayer's mate carried a hundredweight (*c.* 50 kg) of bricks, using a hod on his shoulder, up a ladder. When you think of a Roman building site it is a useful exercise to imagine it without devices like concrete mixers, lifts and power tools – even wheelbarrows!

It is easy to be led by the two main sources of information – Roman authors and modern ones – into believing that there existed some sort of officially sanctioned and right way of doing things. One of the things which have not altered in the building trade is the tendency to improvise. A Roman builder would happily substitute, for example, a roof tile for a brick or a flue tile for a pipe, and a close look at a real building will usually show plenty of '*opus Germanicus*' (archaeologists' modern dog Latin for jerry building) or 'bodging'. How often over the centuries, I wonder, has a building inspector heard those infuriating, mendacious words, 'But we've always done it like that!'

Many years ago I made a study of bath buildings that were attached to Roman domestic sites in Britain. I was an experienced building technologist who had worked *inter alia* on the contemporary use and manufacture of traditional building materials. I had become an expert on the British Standards and Codes of Practice which concerned them. I had a paperback copy of Vitruvius' *Ten Books on Architecture* in my pocket and in my ignorance believed that there must once have been a Roman equivalent of a small-time architect, or perhaps a sanitary engineer, who came along, discussed the proposed baths with the villa owner, and produced a plan from his pocket, with a recommendation for the '5,000-square-foot four-roomer' or whatever.

Imagine my surprise when I found that the 126 suites featured in my study differed widely from one another, not just in detail, but even in plan. Some of them seemed so impractical and exhibited so many later alterations as to suggest that they were almost all either 'do-it-yourself' jobs or else something constructed *ad hoc* by the local bodging builder – if there was such a thing in rural Britain.

The idea of DIY baths leads to a train of thought. Imagine a villa owner in Welwyn (Fig. 58) saying to his agent, 'We ought to have a new suite of baths like the people up the road.' They decide on a site. What happens next? Nowadays the agent might go to B&Q. In my youth we would go to the local builders' merchant. The merchant scribbles on his writing tablet, and comes out with a long list. It included, in the case of Welwyn Roman baths, some 150 *tubuli* (flue tiles – Chapter 22). We know, from the distribution of distinctive roller-stamped flue tiles (not found at Welwyn baths) that either tile makers or their products travelled great distances. In either case, we have two unanswered questions: how did the customer get the materials (or the craftsman) to the villa, and/or how did they get the enormous weight of 'CBM' (Chapter 11) to the site?

I'm afraid I leave answering these questions to someone else. This book is intended to look at Roman building materials and techniques, and the development of the technology of building particularly from around the time the empire was established, when many fundamental changes in materials and techniques were taking place. It also concerns itself with the obvious things that books on architecture seem to take for granted: why do buildings stand up (or sometimes fall down); what sorts of buildings did the Roman build; how did they build them, and what materials did they use?

Most people tend to think that a short event happened, once upon a time, called 'The Roman Period'. The junior school syllabus sees this as a snapshot or a still from a film, or perhaps a collection of insects in a drop of amber. To them 'the Romans' were a small group (all men and all in armour) who somehow conquered and dominated the ignorant natives. It is forgotten that there were millions of Roman citizens, living in an empire which stretched from Scotland to Egypt, and lasted more than one and a half millennia.

The physical evidence discussed in this book depends upon chance survival and must always be seen in its geographical and chronological context. The eruption of Vesuvius in AD 79, for example, was a unique event. It preserved structures, such as wooden window frames and staircases, which had been in place and in use possibly, in some cases, for a very long time; Pompeii and Herculaneum existed for many years before the eruption!

Ostia, which is only is short train journey from Rome, was at one time a town of some 75,000 inhabitants. It began in the fourth century BC and eventually became deserted in the ninth century AD. There is no biodegradable archaeology for the visitor to see, but what impresses are good, solid buildings; literally concrete material of which the potential was only just being realised when Pompeii was buried.

Throughout this book I make use of a treatise on architecture written by a Roman, Vitruvius, but he too represents virtually a moment in time. He was writing before the development and exploitation of concrete, before baths began their conquest of the Empire. Because baths were so influential in the development of architecture (as well as because they were the subject of my research years ago) they are the source of much material in this book.

1

VITRUVIUS

A treatise on Architecture is not like history or poetry. History interests the reader by the various novelties which occur in it; poetry, on the other hand, by its metre, the feet of its verses, the elegant arrangement of the words, the dialogue introduced into it, and the distinct pronunciation of the lines, delighting the sense of the hearer, leads him to the close of the subject without fatigue. This cannot be accomplished in Architectural works, because the terms, which are unavoidably technical, necessarily throw an obscurity over the subject.

These are the daunting words of Vitruvius – Marcus Vitruvius Pollio. He wrote them near the end of the first century BC. Part of his difficulty was that he found writing a painful task. He covers a wide range of topics, from siege engines to astrology, aqueducts to aesthetics. His *magnum opus* is the main – in fact almost the only – contemporary written source on Roman building technology. It must be remembered that, although Roman technology and society were typically conservative, a considerable amount of development occurred after Vitruvius (Fig. 1).

The architect should be equipped with knowledge of many branches of study and various kinds of learning. He ought, therefore to be both naturally gifted and amenable to instruction. Let him be educated, skilful with the pencil, instructed in geometry, know much history, have followed the philosophers with attention, understand music, have some knowledge of medicine, know the opinion of jurists, be acquainted with astronomy and the theory of the heavens.

Nowadays we might say, 'An architect needs to be the Renaissance man.' And, in a way, that came to be true for Vitruvius many hundreds of years after his death, when his *De Architectura* was discovered in the library of the monastery of St Gallen in 1414. It was made widely known by Alberti in 1450 when he used it as a basis for his *De re Aedificatoria*, which was the first theoretical book on the subject to be written in the Italian Renaissance. The first modern edition of Vitruvius' text was published in Rome in 1486.

Vitruvius is invaluable in any study of Roman architecture – so much so, that he is often quoted as if his *Ten Books* were Holy Scripture – but there are questions that should be asked by anyone who reads his work. The first and obvious is: how knowledgeable *was* he? His *curriculum vitae* can only be deduced from what he himself wrote. He was born *c.* 80–70 BC and

Fig. 1 Vitruvius presenting *De Architectura* to Augustus. An imaginary scene from *Vitruvius on Architecture* by Thomas Gordon Smith, 1684.

died *c.* 15 BC. He served in the army as a *ballistarius* (artilleryman). He acknowledges his use of the work of many other authors, but the only actual hands-on experience he tells us of is that he superintended the building of the basilica at Fano, a *colonia* established in Italy by Julius Caesar.

How confident can we be that we understand the terms he uses? The Italians have a saying: '*traduttore? traditore!*' (Translator? Traitor!) The English equivalent of a Latin word in the dictionary might have been decided by a scholar who was ignorant of technical terms used in building. This problem becomes obvious when, for example, in the construction of roofs (Chapter 13) the first citation of a word in a Latin–English dictionary is that the word is used by Vitruvius and was therefore defined by a classical scholar rather than by a builder.

2

BUILDING TYPES

To accustom a population which was scattered and barbarous and therefore inclined towards relaxation and repose by the charms of luxury, Agricola gave private encouragement and public aid to the building of temples, basilicas and houses.

Tacitus, *Agricola* 1, 21

It is easy to get the impression that an unbridgeable gulf separates vulgar 'building' from aristocratic 'architecture'. A book, the title of which contains the word 'architecture', will be likely to concentrate on structures that have survived and which were erected to impress. It might examine in detail their embellishment and superficial ornament while ignoring vernacular, homeward-bound but functional buildings like the privy or the cart shed. In the provinces and away from the towns, different traditions were often followed.

Because it would be impossible to write a book about what buildings were made of, how they were made and how they worked without referring to real structures, a brief introduction to some of the main types of Roman building is required.

The Forum

Forum in Latin is an open space in a town, often a marketplace. In Rome, for example, there was the cattle market, *Forum Boarium*, and the vegetable market, *Forum Holitorium*. Towns always had a formal administrative forum, which was similar to the Greek *agora*, the official meeting place for citizens. It was often surrounded by colonnades; sometimes, as at Pompeii, on two stories. Important temples usually faced into it. Here were the offices of the town officials and the meeting place of the *curia*, and the place of the popular assembly, the *comitium*.

Temples

Templum in Latin originally described a consecrated area in which to look for omens or where sacrifices could be made, but the word 'temple' seems to be used, uncritically, for a wide range of buildings of putative religious use, regardless of their structure. Although the reader probably has some sort of mental picture of what a classical temple *ought* to look like, there is no canonical design. They are like French verbs: nearly all irregular!

A millennium before Vitruvius, it is recorded that Solomon:

> ... built also the House of the forest of Lebanon; the length thereof was an hundred cubits, and the breadth thereof fifty cubits and the height thereof thirty cubits, upon four rows of cedar pillars, with cedar beams upon the pillars. And it was covered with cedar above upon the beams that lay on forty five pillars, fifteen in a row.
>
> 1 Kings 7:1–3

Vitruvius explained Doric architecture in terms of a similar wooden prototype, and this idea is supported, for example, by the fact that the columns of the temple of Hera at Olympia do not match one another; probably because each was replaced in the current fashion as it decayed.

> Following the arrangement of timber framing, workmen have imitated, both in stone and marble, the disposition of timbers in sacred edifices, thinking such a distribution ought to be attended to; because some ancient builders, having laid the beams so that they ran over from the inner face of the walls and projected beyond their external face, filled up the spaces between the beams, and

ornamented the cornices and upper parts with wood-work elegantly wrought. They then cut off the ends of the beams that projected over the external face of the wall, flush with its face; the appearance whereof being unpleasing, they fixed, on the end of each beam so cut, indented tablets, similar to the triglyphs now in use ... so that the ends of the beams in question might not be unpleasant to the eye. Thus the ends of the timbers covered with tablets, indented as just mentioned, gave rise to the triglyph and metopes in the Doric order.

Others subsequently allowed the ends of the rafters above each triglyph to run over, and hollowed out the projecting inferior surface. Thus, from the arrangement of beams, arose the invention of triglyphs; and, from the projection of the rafters, the use of mutules under the corona. On which latter account it is observable, that in works of stone and marble the carving of the mutules is inclined, in imitation of rafters, whose slope is necessary to carry off the water. Hence we have the imitation of the earliest works to account for the Doric triglyph and mutule [Fig. 2].

Vitruvius, IV, ii, 2

The typical Roman temple shows the Etruscan tradition combined with the Greek one. From the Etruscan temple came a high plinth, or *podium*, and frontality, with a columned porch facing east. The façade was usually approached by steep steps. Whereas the sanctum (*naos*) of Greek temples was usually *peripteral* – that is, surrounded by free-standing columns – many Roman temples had half columns engaged at the sides and back of the sanctum (*cella*) instead. This detail supports the cynical jibe that the Romans didn't produce architecture, but created engineering and decorated it with pseudo-Greek details (Plate 1).

Circular temples have a long history. Vitruvius describes the plan of some forms of them. The temple of Vesta in the Imperial forum dates from 715 BC and was one of the most sacred Roman sites. The Pantheon (Plate 2) was a temple.

'Romano-Celtic' temples were constructed in the northern provinces of the Roman Empire. A square shrine was enclosed on all sides by a colonnade, which either stood on the ground or on a breast-high wall. The form is probably pre-Roman (Fig. 3).

In most instances the pagan temple was conceived as an exterior, because it was a 'god box', made to contain the effigy of a deity. The altar, the congregation and the liturgy were outside. Particularly during the later

Building Types

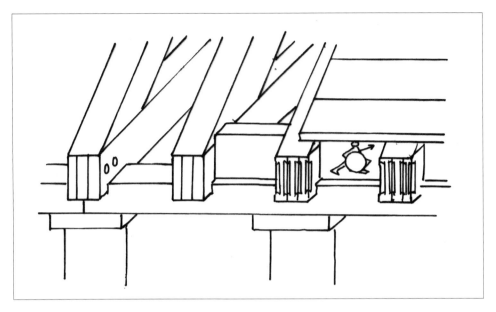

Fig. 2 'Workmen have imitated, both in stone and marble, the disposition of timbers in sacred buildings ... having laid the beams so that they ran over from the inner face of the walls, and projected beyond their external face, filled up the spaces between the beams, and ornamented the cornices and upper parts with elegantly shaped wood-work. They then cut off the ends of the beams that projected over the external face of the wall; ... cut indented tablets, similar to the triglyphs now in use, and painted them with a waxen composition of a blue colour, so that the ends of the beams in question might not be unpleasant to the eye. Thus the ends of the timbers covered with tablets, indented as just mentioned, gave rise to the triglyph and metopa in the Doric order.' Vitruvius IV, ii, 2. This diagram assumes, incorrectly, that the projecting beams were tie beams (Chapter 13).

Fig. 3 A possible reconstruction of a Romano-Celtic temple.

empire, however, there was an influx of 'mystery' religions that were congregational, for which the places of worship were conceived as interiors. For example, Neo-Pythagoreans and worshippers of Mithras met indoors, as did Christians and Jews (although we don't refer to their places of worship as temples). This had an important effect on the history of architecture.

The Basilica

Closely associated with the forum was the *basilica*, a rectangular, aisled hall used as an assembly hall, court of justice and business and money exchange. The first basilica in the forum at Rome was built in 184 BC. It had a high-roofed nave, lit by clerestory windows and separated by colonnades from aisles along the long sides. Opposite its entrance, in an apse, was the tribunal, a raised dais for assessors with an elevated seat for the *praetor* (magistrate). In front of the *praetor* was an altar at which a sacrifice was offered before business was transacted. There were two types: one, with its entrance in one long side, is called 'Vitruvian' (because it was described by Vitruvius, who built one of this plan at Fano), and the other is called 'Pompeian' (from the example at Pompeii) where the entrance is in a short side. The Pompeian basilica became adopted as the Western model for the Christian church, and the name 'basilica' was adopted for churches with other plans (Fig. 4).

A customary building around the Greek agora and at other public spaces was the *stoa*, which was a long, colonnaded portico of a building, structurally (but perhaps not aesthetically) analogous to a modern open-fronted cart or bicycle shed. It is possible that the basilica was developed by combining two *stoas* with a rectangular hall – although this structure only became possible with the introduction of the tie beam (see Chapter 13). Although no Greek example is known, the word *basilica* is a Greek adjective meaning 'royal', which Strabo, around the beginning of the Imperial period, suggests originally qualified the word 'stoa'.

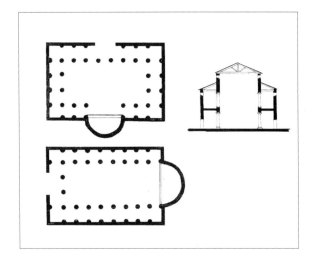

Fig. 4 Plans of the basilica. Top: Vitruvian. Bottom: Pompeian.

Macellum

The *macellum* was a more formal market than the forum. It usually provided shops arranged around a courtyard that contained a central, round kiosk or *tholos* upon a podium reached by steps and had a ring of columns supporting a domed roof. A *macellum* was usually square in shape. The central courtyard of the *macellum* is surrounded by *tabernae* (shops) all of the same size. It was also possible to extend the *macellum* upwards to include upper stories. Entrance to the *macellum* was either through central gates on each of the four sides or through some of the *tabernae* themselves. It appears that the *tabernae* set aside for butchers were together in one area of the *macellum* where they were provided with marble counters, presumably to keep the meat cooler, and drains for the removal of water and fluid waste. It has been suggested that the central *tholos*, also well provided with water and drains, was where fish was sold, or was a shrine to the gods of the market place. It is the presence of this central water feature that seems to denote a *macellum*.

The *macellum* contained fixed standards of length, weight and volume. It is often associated with the forum (Plate 3).

Shops

In Roman towns shops often occupied part of the frontage of town houses. The shopkeeper frequently lived over the shop, sometimes on a mezzanine floor. At Pompeii the bakery, with mills and oven, has a shop-front. There are many *thermopolia* – fast-food shops with *dolia* (large pots) set in the counter for goods. Shops were often associated together in shopping centres. Probably because the development of his new forum displaced small traders, Trajan produced a remarkable shopping complex nearby, which provided at least 150 single-unit shops (Fig. 5).

Fig. 5 Shops on the first floor of the hemicycle of Trajan's Market.

The Theatre

The Greek theatre was about two thirds of a conical bowl (*cavea*) usually carved out of a hillside, with the circular performing space, the *orchestra* (in origin a threshing floor) forming its base. The audience sat in tiers of seats, looking out over the orchestra into an open space. Dramatic presentation changed over time, and part of the action was later transferred onto the roof of a simple building behind the orchestra.

The Roman theatre was similar, but semi-circular and enclosed by a solid architectural construction (*scenae frons*) constructed across its diameter with a raised stage (Fig. 6). The introduction of concrete allowed the Roman theatre, instead of being built into a hillside, to be wholly or partially constructed from the ground up. The *cavea* could also be constructed as an earth bank, stabilised by a timber or masonry revetment.

Fig. 6 The Theatre of Herodes Atticus, Athens, built AD 161.

Roman Building Techniques

Fig. 7 The Colosseum, Rome.

Inset: Fig. 8 The circus of Maxentius. A picture dated 1888.

Building Types

Amphitheatre

An amphitheatre can be looked upon as two theatres joined together (the Greek *amphi* means 'on both sides' – an amphora has a handle on both sides from *amphi* + *phoreus*, a bearer). It was oval with tiers of seats surrounding the flat *arena* where the action, frequently brutal, violent and bloody, such as gladiatorial games, took place. Historically this originally occurred in a convenient open space. Vitruvius says, 'We have inherited from our ancestors the custom of giving gladiatorial games in the forum.' Wooden amphitheatres were built, but none survive. Earthen banks, stabilised by wood or masonry were widespread in the provinces. As with the theatre, the use of concrete made it possible to build amphitheatres from the ground up. The earliest masonry amphitheatre constructed in this way is at Pompeii, which was constructed after 80 BC. The Flavian amphitheatre, known as the Colosseum, is an outstanding and, in many respects, a unique example (Fig. 7).

Imperial amphitheatres comfortably accommodated 40,000–60,000 spectators, or up to 100,000 in the largest venues. They featured multi-storied, arcaded façades and were elaborately decorated in marble and stucco. After the end of gladiatorial games in the third century and of animal killings in the sixth, most amphitheatres fell into disrepair or were demolished.

Circus

The *circus* was a racetrack for chariots, the largest (*Circus Maximus* in Rome) being 650 metres in length. Its structure is analogous to that of the Greek running track *stadion* (Latin *stadium*). Banks of seats lined the long sides and the curved end. At the other end were starting bays (*carceres*). There was a dividing island (*spina*) around which the race was run, along the axis. It was sometimes set slightly obliquely to allow teams more room at the start of a race. Lavish ornaments were arranged along the *spina* (Fig. 8).

Baths

In the Greek world public baths with individual tubs are known from at least the fourth century BC and baths of this type are found in southern Italy. They were incorporated in the early phase of the Stabian Baths at Pompeii. During the Hellenistic period, heated baths became a normal feature of the gymnasium. The Romans replaced hip-baths with shared pools and Vitruvius describes two types of communal bathing establishments. The first were innovative structures dedicated entirely to bathing. The second were those which were annexes to the *palaestra*, the Greek peristyle exercise yard associated with gymnasia.

By Vitruvius' time, the baths had evolved to consist of a series of rooms graduated in temperature from cold (*frigidarium*), through warm (*tepidarium*), to hot and humid (*caldarium*). In addition, early baths included a dry, very hot room (*laconicum*). In public establishments other rooms were added (see Chapter 21).

The invention and evolution of the hypocaust and heated walls, described later, were critical in the evolution of baths, and therefore of architecture. Sir Mortimer Wheeler, in his book *Roman Art and Architecture*, wrote, 'It is an axiom of architectural history that the innumerable public baths of the Roman Empire made an outstanding contribution to the general development of plan and structure.'

Monuments

Statues, busts, inscriptions and columns were erected in large numbers to celebrate achievements and victories. Honorific arches (*fornices*) are known to have existed from at least the second century BC. They might seem to be a peculiarly odd and trivial manifestation in a civilisation that we tend to think of as very practical, but they reminded the traveller of the importance of an emperor, township or divinity and were symbols of the deep superstitions which underlay Roman behaviour, where a doorway was a symbol of the beginning of a journey; the entry into a new land or life.

Triumphal arches were dedicated to an emperor, a member of the imperial family or a general, commemorating a victorious campaign. They also glorified towns or divinities. Their essential feature is at least one vaulted passageway decorated by columns, half-columns or pilasters. An attic usually displayed an inscription and carried statues, trophies etc. Like town gateways, they sometimes have three parallel passageways, the flanking ones being narrower than the central one (Fig. 82). Four-sided arches were occasionally built at crossroads, particularly in North Africa (Plate 4).

Victory pillars were set up to record triumphs. The best known is Trajan's column, up which winds a continuous spiral relief depicting Trajan's campaign in Dacia. Although details of the scenes are to some extent formalised, they are well worth close study (frontispiece).

Tombs

Tombs were the homes of the dead, housed outside the towns, which were the homes of the living. They were, however, more than just monuments. They were places that were visited by the families with offerings for a feast (*agape*) at the festival of *parentalia*. We must remember that only the tombs of the wealthy have survived, and that local, provincial rites existed. One type of tomb was descended from the Etruscan type: a tumulus built on a cylindrical drum. The tombs of Augustus and Hadrian are examples (Plate 5).

Barrel-vaulted chamber tombs, often decorated on the outside, can be seen lining the roads out of towns. They contained niches for the urns of individual cremations, called *columbaria* by modern scholars, from their resemblance to dovecotes. Inhumations became common in the third century AD. Since these needed more space than urns of ashes, they were deposited in *loculi*, which were graves cut horizontally in the sides of tunnels in catacombs, and sarcophagi in chamber tombs.

Military and Defensive Works

Earthworks, such as ditches with banks, possibly topped by palisades and revetted by masonry or timber, are among the earliest and best-known surviving defence works of prehistory. Fortifications for civil settlements

Roman Building Techniques

have a long history. The so-called Servian Wall of Rome dates from 378 BC. It consisted of an earthen bank, faced on the outside by a wall at least 9 metres high, the inner 2.6 metres high. Masonry walls, usually fronted by deep ditches and backed by high banks, were built around towns at many periods, and are difficult to date. It seems that they were often more a symbol of civic pride than a response to any obvious threat.

Permanent military strongholds were divided into larger, 'fortresses' (*castra*), smaller 'forts' (*castella*), and temporary defended accommodation we call 'camps'. The *castrum* at Ostia, which formed the original nucleus of the town, began its life as a defence against sea pirates in the mid-fourth century BC. The usual form of such military sites was playing-card-shaped – rectangular with rounded corners. They were enclosed by a wide ditch in front of an earth rampart (*vallum*) or else a masonry wall backed by an earthen rampart. The defences were pierced by four gateways, and the internal layout was more-or-less standardised (Fig. 9). Many towns began, like Ostia, as a military presence, or with an original rectangular layout set out by the army.

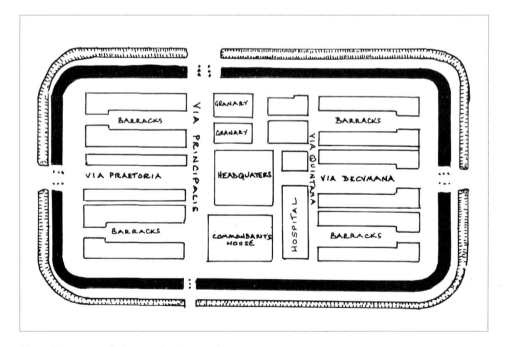

Fig. 9 Diagrammatic layout of a Roman fort.

Roads

For many people roads are among the first manifestations of Roman occupation that spring to mind. However, there are many misunderstandings and prejudices. The search for hypothetical Roman roads is a harmless pursuit for eccentric antiquaries. The strategic or military roads (*viae militares*) are usually obvious, because they cut through the landscape, often in straight lines, diverting only for natural obstacles like steep hills. There were state roads (*viae publicae*), which are well known, and bear the names of their sponsors. Other classes, according to the first-century *agrimensor* (land surveyor) Siculus Flaccus, were the *actus*, built and maintained by the local tribes, and private or estate roads.

The structure of roads was not (as some textbooks would have us believe) prescribed; Vitruvius' careless remark about building materials – 'use them as you find them' – might have been made about road metalling, even of military roads, especially in the provinces. The road from Chichester to London is, in part, built using slag from the Wealden ironworks – metalling indeed! The main road from Verulamium (near St Albans) to Colchester, the route of which seems obvious on the map and on the ground, was metalled in flint at some points, but is impossible to detect where it crosses gravel. The road's important features included side ditches, which provided material for a raised embankment (*agger*), good drainage and a delineated military space, which can be seen on some air photos.

Bridges

Roads crossed rivers by fords, often paved, or on bridges. Timber bridges are known from the description by Caesar (Chapter 4) and other literary sources, and from representations, for example one with masonry supports shown on Trajan's column, which also shows a pontoon bridge. Vitruvius tells how to use a coffer dam (V, xii) or a caisson (X, iv & vi) to enable work to go on in a river. Masonry bridges, of course, made use of arches (Plate 6), which are discussed elsewhere. The first stone bridge in Rome was built with a wooden superstructure in 179 BC. Masonry arches were added in 142 BC.

Aqueducts

Water was provided to some towns from great distances by carefully surveyed channels, and the rare but spectacular multi-arched aqueducts are symbolic of Roman civilisation (Plate 16). Since an aqueduct flows continuously, it also provided for the flushing of sewers, which were often also remarkable, but not so visible. Water to most places was provided by wells (Chapter 19).

Latrines

A student of life in almost any period in the past will almost always find himself faced with the question, 'where did they relieve themselves?' Just imagine: the palace at Versailles has no lavatory! There are very few surviving Roman public latrines. There were rows of seats around the walls, side by side over a deep channel containing flowing water. In front of them was a water-filled channel in the floor for rinsing the mops, such as sponges (or the like) on sticks, used as the equivalent of toilet paper. Men and women did not have separate lavatories. Roman costume, which in general had long skirts, ensured that privacy was maintained because there were openings in the front of the toilet seats which permitted use of the sponges on sticks whilst remaining seated, since the skirts of their garment reached their knees (Fig. 10).

Fig. 10 The public lavatory at the Forum Baths, Ostia.

Building Types

Domestic Buildings

For the purposes of this book it is not necessary to attempt to classify or describe the many different types and plans of dwelling that existed over the broad time-span and wide geographical distribution. They range from town houses and apartment blocks and from palaces and luxury villas to small, vernacular farmsteads and hovels.

3

SURVEYING

Surveying is a precondition for building. The principal processes involved in simple surveying are the measurements of angles, distances, and levels. It is surprising that the ways in which the Romans measured distances, particularly longer distances, are poorly documented. Heron, in the first century AD, refers to a 'band' or a chain for this purpose. It seems unlikely that the use of carefully tensioned ropes, which has been suggested, is sufficient to account for the accuracy claimed by modern writers.

One suggestion was the use of a system that counts the number of revolutions of a wheel, but nothing survives which is anything like a simple wheeled hodometer, and in any case such devices would not be accurate over rough ground. It is questionable whether there *were* simple measuring wheels, since all the written accounts are of ambitious devices with gears, suggesting that perhaps the writers were displaying a false ingenuity. That described by Vitruvius (X, ix, 1–6) in long-winded, tedious detail, would record the distance by dropping pebbles into a container. It seems highly improbable and unlikely to have been constructed, and should be dismissed as an 'armchair' invention. He assumes the circumference of a wheel of diameter 4 feet to be 12.5 feet: a systematic error of 0.55 per cent. There is some doubt about the accuracy of the existing text of the description of the hodometer of Heron (*Dioptra*, Chapter 24). He specifies wheels by their circumference, not saying how they would be created. An account by Pliny (*Natural History* 6, 61–2) does make it seem likely that the bematists (pace counters, from the Greek *bema*, a pace) of Alexander the Great might have used such a device.

To locate a point on a flat surface usually relied on making two linear measurements: one along a straight datum or base line (which could easily

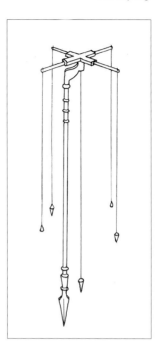

Fig. 11 A reconstruction of a *groma*, based on the specimen from Pompeii.

be set out using a stretched string or optically by aligning poles) and one (an offset) at right angles to the first. A right angle is easy to measure by making a triangle, the sides of which are multiples of 3, 4, and 5 units, or using a cross-head staff. This would have consisted of a rod on top of which was mounted a box with slots in the sides, accurately aligned at right angles. It is mentioned by Aristophanes in *The Birds* and was employed by Meton.

The best-known Roman surveying instrument was the *groma*, an example of which was found at Pompeii. The instrument is described by Frontinus in his work on aqueducts, which was written in the late first century AD. It consisted of a cross, from the ends of which plumb lines were suspended, which was mounted on a bracket from the top of a staff. Both the cross and the bracket were pivoted to rotate in a horizontal plane. A plumb line hung from the centre of the cross. In use, the staff was located close to the base line and made vertical by aligning a pair of plumb lines against it. The bracket was then turned until the centre plumb line was over the base line, and the cross was rotated so that one pair of plumb lines was aligned by eye along it (Fig. 11).

Vitruvius (VIII, v) mentions, but does not describe, the *dioptra*, which permitted measurement of angles other than right angles. No example or

Roman Building Techniques

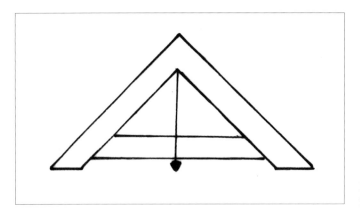

Fig. 12 A combined level, plumb bob and square.

contemporary illustration of this instrument is known, but it was probably a circular plane table, with a scale (marked in degrees?) around the circumference, and with the *alidade* (sights) pivoted at its centre. A more sophisticated version, improved to measure angles, is described by Meton. He liked gears, and this device makes use of cogs and worm gears to pivot an *alidade* in both horizontal and vertical planes. How it actually *measured* angles is not known. It sounds like a theodolite without a telescope.

Vitruvius also lists a water level, which was effectively a long U-tube. It consisted of a horizontal metal pipe on a stand, terminating at either end in upstanding glass tubes. The *chorobates*, which Vitruvius preferred, consisted of a length of wood with sights. (Vitruvius suggests it should be as long as 9 metres!) It was levelled by plumb lines hanging beside the legs of a stand. In addition there was a trough on the top, which could be filled to the brim with water – a sort of inverted 'spirit level'. The modern spirit level, which makes use of a bubble in a curved tube or convex glass, was not known, but a number of levelling devices that made use of a plumb line have been found. They mostly doubled up as set squares (Fig. 12).

Vitruvius describes how to orientate streets and buildings by tracking the shadow of a gnomon, marking lines when it is of equal length, and bisecting the angle between them to mark south.

4

FOUNDATIONS

It is common to come across the statement in British archaeology that 'the foundations of a Roman villa were discovered...' What is usually found are the bases of walls, which can consist of three layers: the foundations, the footings, and the bottom of the wall proper. These terms are best illustrated by diagrammatic sections (Fig. 13).

To create foundations a trench was dug and sometimes lined with timber shuttering. Suitable material, such as concrete, gravel, or stone was rammed into it. On the foundation, broad *footings* were constructed and then the wall was built. On many Romano-British villa sites the footing was, in fact, the plinth on which the ground sill (the bottom beam) of a timber frame was laid. The footings under the columns of a temple were in the form of walls called *stylobates* (Fig. 14).

It seems that many people don't understand what foundations do, since, in recent years, some archaeologists have dreamed up a nonsense idea that there must be a correlation between the depth of foundations and the height of the structure they support. The purpose of foundations is to spread the weight of the structure into the subsoil. What is required depends on the nature of the subsoil and, until modern science, has always been a matter of judgment. Vitruvius (III, iv) says, 'the foundations ... should be dug out of the solid ground, if it can be found, and carried down into the solid ground as far as the magnitude of the work shall seem to require'.

This apparently rather vague instruction was more or less repeated by textbooks nearly 2,000 years later, pointing out the value of experience in deciding how deep to go. One nineteenth-century book succinctly tells us,

Roman Building Techniques

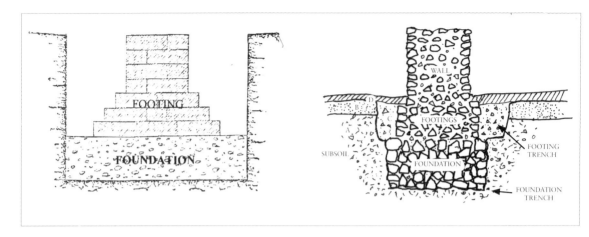

Above: Fig. 13 Terms used for the construction of the bases of walls. The section on the right is imaginary. Archaeologists (and others) commonly call all the structure below the original ground level 'foundations'. The footings trench C was dug after layer B was deposited, and was sealed by layer D. It could be dated by finds from B.

Left: Fig. 14 'Walls should be constructed under the columns half again as thick as them ... Called "ground walkers" (stereobates) ... The spaces between them should be vaulted over or rammed with fill in order to stabilise them.' The footings of the Temple of Hadrian in Rome do not exactly fulfil the instruction in Vitruvius III, iv. The material visible between the stylobates (which usually run *along* the lines of columns) is brick-faced concrete or dressed stone.

Foundations

A good foundation should fulfill the following conditions:

1. It must be incompressible, or at least equally yielding throughout.
2. It should be perpendicular to the pressure applied to it.
3. It should be of sufficient area to bear that pressure.
4. It should be unalterable in nature, either by atmospheric or other influences.

Vitruvius adds, 'If, however, solid ground cannot be found … it must be dug up and cleared and set with piles made of charred alder, olive or oak and these must be driven down by machinery, very close together, like bridge piles …'

Charring is said to harden and preserve the surface of timber. It can, to some extent, sharpen piles, by burning them to a point.

When Vitruvius was young, Julius Caesar built a bridge across the Rhine supported on piles:

> He joined together at the distance of two feet, two piles, each a foot and a half thick, sharpened a little at the lower end, and proportioned in length, to the depth of the river. After he had, by means of engines, sunk these into the river, and fixed them at the bottom, and then driven them in with rammers, not quite perpendicularly, like a stake, but bending forward and sloping, so as to incline in the direction of the current of the river …
>
> Caesar, *The Gallic Wars*, Book V

Unfortunately we have no information about either Caesar's or Vitruvius' pile-driving machinery. Probably they used a pile driver ('crab'). A heavy weight ('monkey') was repeatedly wound up by a windlass on shearlegs (similar to Fig. 23) and dropped to hit the tops of the piles.

Unless the sharpened tips of the piles struck solid rock there would, of course, be no definite limit to their penetration, and a formula was used to make the decision when to stop. It connected the weight of the monkey, the height it fell and specified the distance to be driven by the last blow required, as 'Piles shall be driven until successive blows by a 200lb monkey produce less than 1/8 inch further penetration'. The Roman engineers must have developed some similar rule of thumb.

5

WOOD

The carpenter stretcheth out his rule; he marketh it out with a line; he fitteth it with planes; he marketh it out with the compass...

Isaiah 44:13

Wood has, of course, been used for construction from prehistoric times. The surviving remains of Roman carpentry and the Roman carpenter's tools seem surprisingly similar to modern ones (Fig. 15). What *has* changed recently has been the introduction of portable power tools, and the provision of more efficiently tempered steel tools.

There is a tendency for us to think that in Roman Britain wood was an inexhaustible commodity, both for use as timber and as fuel. However, when we consider the demands of, for example, the iron industry, it becomes obvious that, had the charcoal burners clear-felled the forests, they would have run out of raw material in a generation. Forests must have been managed by coppicing.

Coppicing (the English word comes from the Norman French '*copiez*', cut) relies on the fact that, after a deciduous tree is cut at its base, new shoots spring up around its perimeter. These grow into poles and, if permitted, become small trees. If at some stage in they are felled, the regeneration starts again, and the process can be repeated. Some coppice 'stools', as they are called, must be many hundreds of years old and many feet across. 'Standards' – trees which are not coppiced – were usually allowed to grow to maturity in a coppice before they were felled as timber. The hypocaust of even a small bathing suite would have required the output of several hectares of coppice. In AD 354 the people of Rome helped to relieve a famine at Terracina, not

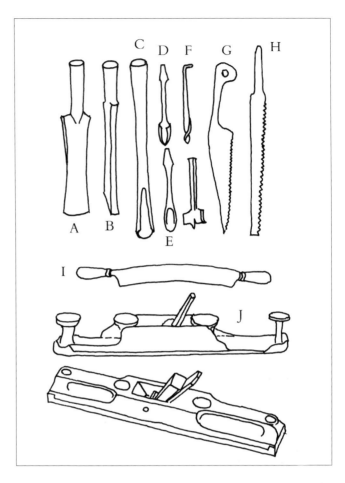

Fig. 15 Woodworker's tools:
A: socketed chisel (London),
B: morticing chisel (Silchester),
C: socketed gouge (Silchester),
D: spoon bit (Zurich Museum),
E: gouge bit (Bucklebury),
F: centre bit (Zurich Museum),
G: saw blade (London),
H: saw blade (London),
I: spokeshave or draw knife (London),
J: The iron-shod plane (from Verulamium) is a surprisingly common type. A restoration is depicted at the bottom.

from benevolent motives, but because Terracina supplied wood to fuel the huge baths in Rome, 70 miles away.

Timber for building was obtained by felling mature trees. This could have been achieved using only an axe, but remains of two-handed saws are known, and it is likely that the Romans employed the method common recently, before the advent of powered chainsaws: wedge-shaped notches were cut from either side of the trunk as close as possible to the root with a deeper one on the side to which the tree will fall. The saw is inserted into the shallower cut and operated by a man on either side. Wedges were driven in behind the saw as it progressed, to keep the cut open and to ensure that the tree fell in the desired direction. Branches were removed from the fallen tree by the axe or billhook.

Roman Building Techniques

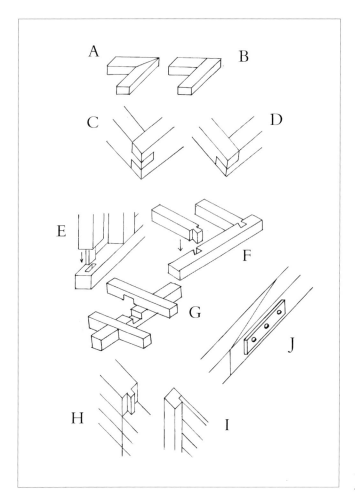

Fig. 16 Some examples of Roman woodworking joints: A: mitre, B: butt, C: bridle, D: half lap (corner halving), E: mortice and tenon, F: dovetail, G: cross halved, H: tongue and groove, I: rabbet (rebate), J: scarf.

The bark from an oak tree was used for tanning leather and, as in more recent times, was probably as valuable as the timber. After it had been removed the trunk was split into usable planks or beams which were shaped using an adze or a broadside axe (one with the blade offset). A more labour-intensive method was to use a 'rip' saw (one set to cut along the grain). This was normally a frame saw. Numerous pictures and reliefs show this process; the timber is supported on trestles. No Roman saw pit has been recognised.

Availability, weight, ease of transport and of fabrication and permanence were obvious considerations. Otherwise the behaviour of the material under load was (and still is) of extreme importance. From this point of view stone, brick and concrete, although they show considerable variation, are rigid

Wood

and strong under compressive load. A considerable force is needed to crush them. However, they are weak in tension; they break if you pull them apart (Chapter 25). Wood, on the other hand is flexible, relatively weak under compression, but strong in tension *along the grain.*

Impressions of timber are occasionally found in, for example, mortar or fired clay. Otherwise, except when it has been buried in contact with a preservative such as copper or lead, wood only survives under extremely wet or extremely dry conditions, or when it has become carbonised by heat in the absence of air, as at Herculaneum. This must be taken into consideration when we think of Roman buildings, in which we have to consider the use of biodegradable materials. An unusual case is the preservation of the void left by vanished timber which has been cast in plaster by archaeologists. This has proved very useful when it comes to the study of doors in the cities around Vesuvius (Fig. 49).

A modern joiner would recognise the tools and the joints used by his Roman counterpart (Fig. 16). We know, however, that even relatively recently there has been an evolution of techniques and changes in fashion in the construction of timber-framed buildings. The modern expert on surviving upstanding structures will sometimes date them by the joints employed, particularly in the case of scarf joints, which are those employed to join pieces of wood lengthwise. He would probably find it difficult to create a chronological sequence for the Roman period because of the paucity of surviving specimens.

6

STONE

Apart from improvements in the material from which tools were made, the hand tools that were used by the Romans to work stone were much the same as those used for hundreds of years before and most of which are still in use today (Fig. 17).

Water-powered stone saws are mentioned by Ausonius in the poem 'Mosella' in the late fourth century AD. Writing of the River Erubius (modern Rewar) he tells us:

> He, turning the millstones with rapid whirling motion And drawing the screeching saws through the smooth white stone Listens to an endless uproar from each of his banks.
>
> 'Mosella', lines 361–64

Mechanical cutting of fine stone, such as marble, appears on the third-century sarcophagus of Aurelius Ammianus, a miller, at Hierapolis, which shows what is interpreted as a watermill cutting stone. It seems to show frame saws being operated by cranks, which are otherwise not known at this period. The pump described by Ctesibius to operate an organ was driven by a projecting rod, acting like a cam, but it relied on gravity for the return stroke (Figs 18 & 19).

Typically, the Romans made use of whatever stone was available locally. For the construction of Londinium, ragstone was brought by barge from Kent. Some prestige projects required special stone which was transported considerable distances. For example, at Leptis Magna (now in Libya), marble was used which came from all round the Mediterranean.

Stone

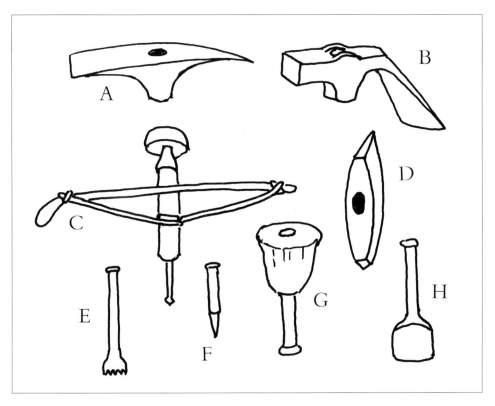

Fig. 17 Masons' tools: A: chisel-ended pick, B: dressing hammer, C: bow drill, D: scabbling (or sappling) hammer, E: claw chisel, F: mason's point, G: mallet, H; drove, nicker or bolster. All of these are common Roman types. The bow drill is depicted on funeral monuments but, like the wooden mallet, only rarely survives except in very dry or waterlogged conditions.

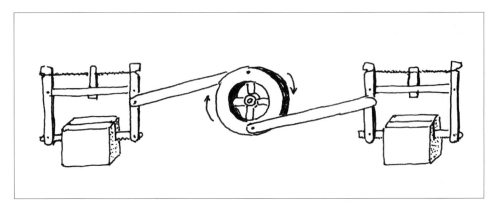

Fig. 18 A possible interpretation of the water-driven sawmill depicted on the sarcophagus of Aurelius Ammianus, at Hierapolis. The central wheel is driven through gears from a water wheel. This reconstruction implies the use of a crank, which is usually denied by writers on Roman technology.

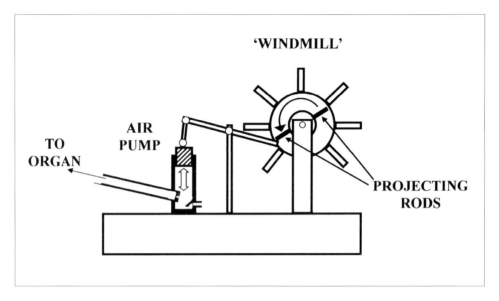

Fig. 19 A possible reconstruction of Ctesibius' wind-driven organ. The lever is operated by projecting rods or spokes, and the return movement was produced by gravity. It is generally believed that the crank was unknown to the Romans. The mechanism shown here, however, could not have been used to work the stone saw shown in Fig. 24.

Freestone was obtained by removing a series of steps, in open quarries or in tunnels. Vertical slots were cut to form the ends of blocks, and use was made of wedges in slots made behind and under it to release it from the quarry face. Columns and column drums were usually cut in the round horizontally (Fig. 20) but some drums have been found still *in situ* that were cut round vertically.

Stone was manoeuvred by levers and pulleys and scanty remains of wooden pulleys have been found. Blocks that show handling bosses, and holes for both pincer grips and lewises (Figs 21 & 22) indicate the use of cranes. In Book X, chapter ii, Vitruvius describes three types of crane (Fig. 23). The first is the *recamus*: shear legs steadied by guy ropes and operated by a windlass and using pulley blocks. For large loads he recommends heavier shearlegs erected by a stay rope and pulleys passing to another windlass held in place by piles driven into the ground. Then two sets of lifting pulleys should be used, and, in effect, two windlasses in series. The first takes up the rope from the pulleys, and is turned by a drum, which is turned by the operators. He then says that the drum can be made large enough to become a treadmill. The

Stone

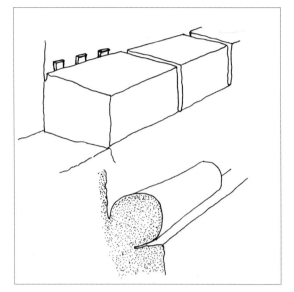

Right: Fig. 20 Common quarrying techniques. Top: removing rectangular blocks with wedges. Bottom: fabricating undressed columns *in situ*.

Below: Fig. 21 Lifting a block of stone. A: sling around lugs which are dressed off *in situ*; B: Rope sling threaded through a purpose-cut loop; C: calliper tongs or nippers, D: chain lewis; E: three-legged lewis; F: the assembly of a modern three-legged lewis.

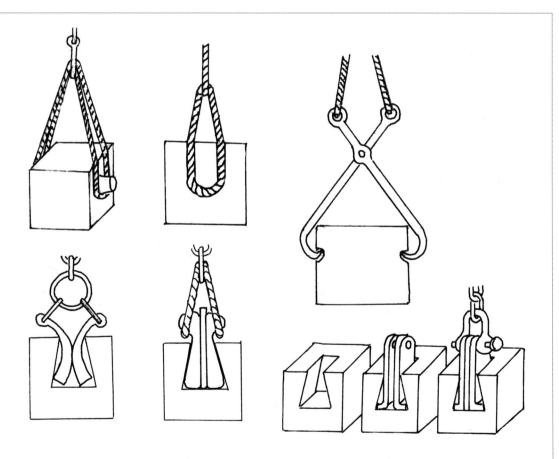

Roman Building Techniques

Fig. 22 The abutment of the bridge carrying Hadrian's Wall over the North Tyne at Chesters, the stone of which shows tool marks from dressing, lewis holes and grooves for tie rods.

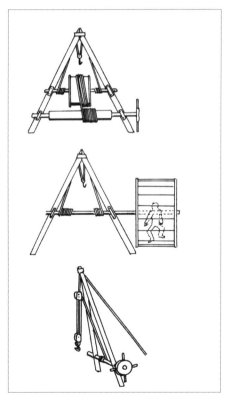

Fig. 23 Cranes described by Vitruvius. The upper and lower windlasses are operated by capstans; the centre one by a man, or men, in a treadmill as in Fig. 24.

Stone

Fig. 24 Monument of the Haterii, in the Vatican Museum, showing a treadmill crane with a quintuple-pulley system. At the top two men are fastening a rope, perhaps round a fender to prevent damage to the masonry. The shear legs are erected and supported by guy ropes operated by pulleys, second century AD.

tomb of the Haterii, now in the Vatican museum, illustrates this (Fig. 24). He also describes a crane with a single-beam 'jib' pivoting on a vertical mast, and mentions modifications of cranes for other purposes. In the next chapter he explains the physics of the pulley with a windlass or capstan.

How the larger, roughed-out blocks, columns and drums were transported is often the subject of discussion and speculation. The most obvious ways are rollers, sledges or carts, perhaps running on temporary tramways. Vitruvius tells us of two ingenious methods, one used by Chersiphron, the architect of the temple of Diana, which turned column drums into rollers, and another, by Chersiphron's son Metagenes, who created solid wooden wheels round the ends of architrave blocks. In each case oxen could have provided traction. A modern parallel was the transport of a 600-tonne monolith from the marble quarries at Carrara to Rome for Mussolini's Olympic Stadium in 1928, on a sledge, using sixty oxen.

7

MUD

Give 'er a good 'at and a good pair o' boots, and she'll last a lifetime.

Advice for building a cob wall

In other words, 'keep the rain out of the top and the rising damp from the bottom'. You can see this taken seriously in the West Country, in cob-built cottages with wide thatched eaves and brick foundations. A similar provision would be usual in any building constructed in unfired clay.

Cob (also known 'clob', 'clem' or 'cleam' in local dialect – it is a vernacular building technique) is made by mixing mud, made to a suitable consistency – with fibrous organic material, such as grass, chopped straw or reeds – and compacting it by treading to make walls, without shuttering or formwork. The walls are trimmed with a paring iron like a Dutch hoe and allowed to dry before being whitewashed.

Similar walls, known as *pisé de terre*, make use of shuttering made of wood or basketwork. Pliny (NH 25, 11, 8) refers to walls made 'by enclosing earth in a frame of boards'. A corridor villa in Hertfordshire produced no evidence of internal walls; they must have been of mud that had washed away. At Lullingstone Roman villa, however, the excavators were alerted to the existence of a *pisé* wall by the sight of two parallel lines of plaster in the mud, which represented the faces of a collapsed wall. Wheeler tells the story of an archaeologist in the Middle East who excavated the walls, leaving the rooms standing!

Adobe walls are built with hand-moulded earth or of 'clay lumps', stuck together with wet clay, which is then used as a finish coat before whitewashing.

Mud

Archaeologists use the anachronistic term 'bricks' for building components made by moulding plastic mud into similar shapes by hand. They were sun-dried and then laid in mud 'mortar'. Cigar-shaped and hog-backed bricks were used at Jericho in the pre-pottery Neolithic at least as early as the sixth millennium BC. More sophisticated mud bricks, cast in rectangular wooden moulds, were used in Egypt for thousands of years, but are rarely seen being made today except on ancient monument sites, since the construction of the Aswan dam has stopped the inundation and the resulting regular supply, of mud, and subsequent legislation prevents the use of the existing supply.

Visitors to Roman sites in Britain often see walls a couple of feet or so high. They might be the bases of masonry walls which have been demolished, but frequently they were the plinth of a timber-framed wattle-and-daub or mud wall – the 'good pair of boots', in fact. The 'good 'at' was provided by the wide eaves of the roof.

It was once a common misconception that the pre-Roman Britons were primitive and lived in mud huts. In fact, their culture was anything but primitive and although their dwellings may have been (partially, at least) made of mud, they weren't necessarily huts (Fig. 25)! In fact most of us really live in mud houses. Take the traditional mid-twentieth-century English 'semi', for example. Its walls are made of bricks (made from clay, which is a mud that hardens in the fire) and set in mortar (another sort of mud) on foundations of concrete (a special mud that sets hard). Its roof is made of tiles (clay, fired in a kiln). The walls are plastered (another special mud) on the inside, and perhaps rendered on the outside (with a mud similar to mortar).

The walls of many of the pre-Roman British round-houses, and of many of their Romano-British rectangular successors, were of 'wattle-and-daub'. A timber frame was erected to support wattles, which were basketwork woven of suitable flexible sticks such as willow or hazel wands. Mud was plastered on to completely enclose the wattles. The essential ingredient of material used was clay, with water which imparts plasticity and coherence. When clay dries, however, it shrinks, and this causes cracking. The builder either chose a source that naturally contained a suitable tempering agent, such as sand or chalk, or added it. Chopped or broken straw was often used, as it not only modified the plasticity of the clay, but also assisted drying from the interior and distributed cracking. The mixture was 'puddled' or 'pugged' with water

Roman Building Techniques

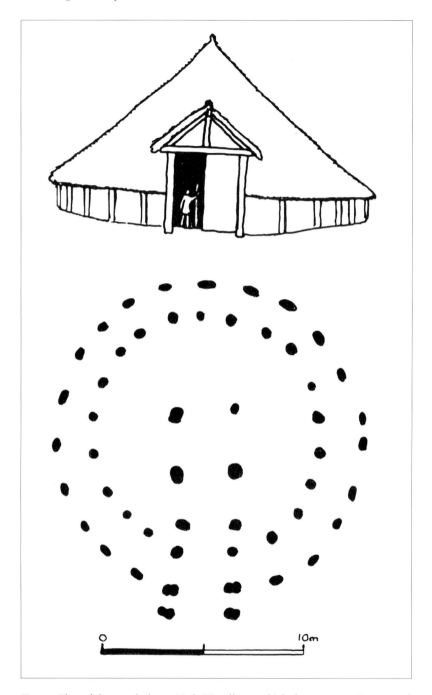

Fig. 25 Plan of the post holes at Little Woodbury, which the excavator interpreted as evidence for an Iron Age 'hut'. The elevation is a suggested reconstruction. The building is about 17 metres across and 10 metres high. Its ground plan area of 226 square metres should be compared with the *total* floor area of a pre-war 'semi' of about 150 square metres.

Mud

to give suitable consistency. This practice, which Vitruvius merely mentions, was considered essential by the Hebrews in the time of Moses.

Vitruvius (II, viii, 20) disliked this method of construction because

> It makes cracks from the inside by the arrangement of its timbers for these swell with moisture as they are daubed, and then contract as they dry, and solid plaster is split by their shrinking. The more it saves in time and gains in space, the more general is the disaster it may cause; for it is made to catch fire like torches.

This ancient catastrophe might have proved a bonus to some future archaeologist, as daub can fire to the consistency of brick, preserving evidence that would otherwise be destroyed by the weather (Plate 7). More often, however, where the daub has a low clay content, and having been fired terracotta red, it subsequently breaks up. Archaeologists call all small, brick-red, soft fragments 'daub', regardless of their origin. Unless they *were* destroyed by fire, structures of dried mud didn't usually survive the weather in Britain.

Possibly one of the earliest and most primitive construction methods, both for small domestic buildings and for large earthworks, was using turves, cut in manageable sizes. This was a method extensively used by the Roman army for works from the ramparts of temporary camps to the massive frontier works, such as the Antonine Wall and part of Hadrian's Wall. The Roman military author Vegetius (*De Re Mil*, iii, 8) prescribes standard dimensions for turves for ramparts as 18 inches (46 centimetres) long, 12 inches (30 centimetres) broad and 6 inches (15 centimetres) thick.

8

LIME

When they are described, some processes seem laughingly and illogically unlikely. Making a slice of buttered toast is simple, but imagine explaining to an extraterrestrial visitor how it is produced: starting from a cow eating grass in one field, and the seeds from a different sort of grass growing in another! What about spreading some marmalade on it? At first, making lime mortar also sounds simple. All you needed to do is mix sand – a universally available material – with lime. But how did you make lime? You *burn* rocks!

Limestones are naturally occurring types of rock, mainly calcium carbonate. They vary from very soft stone like chalk, to strong building stone. To make lime, it was burnt, usually where it was quarried, in a more-or-less cylindrical kiln. Two main types of kiln are possible: the flare (periodic or intermittent) kiln, and the draw (continuous or running) kiln. In the continuous kiln the fuel was mixed with the limestone and put in the top and quicklime was drawn from the bottom. In the flare kiln, which was the usual Roman type, the limestone was carefully packed, creating an improvised corbelled dome of the limestone at the bottom which left a space in which the fuel was burnt.

Cato, writing in about 160 BC, describes what is clearly a flare kiln:

Build the lime-kiln ten feet across, twenty feet from top to bottom, sloping the sides in to a width of three feet at the top. If you burn with only one door, make a pit inside large enough to hold the ashes, so that it will not be necessary to clear them out. Be careful in the construction of the kiln; see that the fireplace covers the entire bottom of the kiln. If you burn with two doors there will be no need of a pit; when it becomes necessary to take out the ashes, clear through one door while the fire is in the other. Be careful to keep the fire burning constantly, and do not let it die down at night or at any other time. Charge the kiln only with good stone, as white and uniform as possible.

In building the kiln, let the throat run straight down. When you have dug deep enough, site the kiln so as to give to it the greatest possible depth and the least exposure to the wind. If you lack a spot for building a kiln of sufficient depth, run up the top with brick, or face the top on the outside with field stone set in mortar. If the flame comes out at any point but the circular top when it is fired, stop the orifice with mortar. Keep the wind, and especially the south wind, from reaching the door. The calcining of the stones at the top will show that the whole has calcined; also, the calcined stones at the bottom will settle, and the flame will be less smoky when it comes out. (Fig. 26)

De Agri Cultura, 38

When heated to about 900°C for some time, limestones lose about 40 per cent of their weight and become what is known as quicklime.

Fig. 26 A flare lime kiln according to Cato.

$$CaCO_3 = CaO + CO_2$$

Quicklime reacts spectacularly with water, becoming hot, producing clouds of steam and crumbling into a fine powder if the quantity of water is limited.

$$CaO + H_2O = Ca(OH)_2$$

This powder, slaked lime, mixes with excess water to form 'lime putty'.

Occasionally, as at Pompeii when repairs were being made after the earthquake of AD 62, small kilns were constructed on site. It is doubtful if these could reach the required temperature to make quicklime. They were probably used to make gypsum plaster (Chapter 10). Quicklime was usually delivered to the building site, where it was slaked (or 'slacked') in a pit lined with boards. If the temperature in the kiln had been too hot, it is said that some of the quicklime was 'overburnt'. Particles could hydrate slowly, causing pitting of a finished plaster surface. The putty was therefore left for days to mature; Pliny (NH 34, 23, 55) specified a period of three years! It was mixed with a larry: a tool similar to a gardener's chop hoe or a lightweight adze (Fig. 27A).

> When the slaking is properly conducted, and care taken in the preparation of the materials, a hoe is used in a way similar to that with which timber is hewn, and the lime is to be chopped with it, as it lies in the heap. If the hoe strikes upon lumps, the lime is not sufficiently slaked, and when the iron of the instrument is drawn out dry and clean, it shows that the lime is poor and weak; but if, when extracted, the iron exhibits a glutinous substance adhering to it, that not only indicates the richness and thorough slaking of the lime, but also shows that it has been well tempered.
>
> Vitruvius, VII, ii, 2

Historically, lime putty was probably first used in building diluted as white paint – a 'lime-wash'. It is likely that it was later used as a lubricant, enabling stone blocks to be slid into place. Later, mixed with sand, it made a sort of mud with special rheological (flow) properties. It could be spread with a trowel, but then stayed put. In other words, it was thixotropic, like modern 'non-drip' paints.

9

MORTAR & CONCRETE

Vitruvius (I, v, 8) told the Romans to build walls of 'freestone, flint, rubble, burnt or unburnt bricks – use them as you find them'. When walls were made of rough stone or brick (either fired or sun-dried), some sort of mud was needed to fill the spaces between them. By the end of the third century BC the 'mud' that was fundamental to Roman building was mortar. Vitruvius (II, v, 1) describes how to make mortar. Like modern builders, the Roman ones measured quantities by volume.

> Three parts of sand are mixed with one of lime. If river or sea sand be made use of, two parts of sand are given to one of lime, [putty?] which will be found a proper proportion. If to river or sea sand, potsherds ground and passed through a sieve, in the proportion of one third part, be added, the mortar will be better for use.

Bricklayers say, teasingly, 'mortar is used to hold bricks *apart*'. Lime mortar is not made, as you might think, to stick them together. It is spread to make it easy to place the bricks and then it stays put. It dries slowly so that, as work progresses, more bricks can be added on top without it being squeezed out. When dry it is a very weak solid, but it doesn't need strength; its function is to smooth out the irregularities of the faces of the bricks so that the load of the building further up is evenly distributed (Chapter 25). Lime mortar doesn't 'set' in the usual sense of the word although, properly used, it dries sufficiently to allow successive layers of masonry to be constructed.

The use of squared (ashlar) masonry was not economical, especially in areas where suitable freestone was not available. Successive thin layers of mortar could be spread, and rough, unshaped pieces of stone (*caementa*) were laid in them. This provided *opus caementicium*. It was not poured, neither did it set. Otherwise, it is what we, today, might call 'concrete'. In the early nineteenth century the word 'concrete', the antonym of 'abstract', came to be used to describe the mixture of mortar and aggregate. Just as with the word 'brick' (Chapter 11), there is a confusion between modern terminology and ancient practice. In the past 500 years the English word 'cement' (from *caementa*) has come to mean, not the aggregate, but a material which, mixed with water, sticks the aggregate together and then hardens irreversibly. The Romans did have a naturally occurring source of such a material: *pulvis puteolanus*.

> There is a species of sand which, naturally, possesses extraordinary qualities. It is found about Baiæ and the territory in the neighbourhood of Mount Vesuvius; if mixed with lime and rubble, it hardens as well under water as in ordinary buildings.
>
> Vitruvius, II, vi, 3

This volcanic ash, named after Puteoli (modern Pozzuoli) in the region north of Naples, was used where a mortar was required which set or, more importantly, was waterproof, as in the linings of aqueducts. It was used only where it could be brought economically. Similar 'pozzolanic' materials did occur elsewhere and were exploited by the Romans: 'trass' from Eifel, the Moselle, Nette and Brohl valleys was used for Roman waterworks in Gaul. Crushed tile with lime produced a 'pozzolanic' set, and a pink concrete, known to archaeologists as *opus signinum*, was also used in situations where resistance to damp was important, as in baths, tanks, aqueducts and floors.

There is a common myth about Roman masonry: that there was some secret, almost magic formula, now lost. The survival of most of the ruins we see today is simply because they have been there for a very long time. The lime in the mortar has quietly absorbed carbon dioxide from the air and once again become limestone. The process of calcining has been slowly reversed:

$$CaO + CO_2 = CaCO_3$$

Mortar & Concrete

A product that was called 'Roman cement' was invented in 1796. It was made by burning septarian nodules, which were washed out of London clay by the sea along the East Anglian coast. It is ironic that there are the remains of Roman buildings in which septarian nodules were set in lime mortar!

Modern mortar and concrete usually contain 'Ordinary Portland Cement' (OPC), which was developed in 1845. It is a fine powder produced by grinding a clinker formed by fusing a mixture of limestone or chalk with clay at around 1,450°C and adding a small quantity of calcium sulphate (plaster of Paris) to control the rate at which it sets.

10

PLASTER STUCCO & FRESCO

Vitruvius tells us about plastering a wall:

> The first coat on the wall is to be laid on very roughly, and, while it is drying, the sand coat, setting it out, in the direction of the length, by the rule and square; in that of the height, perpendicularly; and in respect of the angles perfectly square; so that the finished plastering will be true for the reception of paintings. When the work has dried, a second and afterwards a third coat is laid on. The sounder the sand coat is, the more durable the work will be.
>
> When at least three more sand coats have been applied, the coat of marble-dust follows; and this is prepared, that it does not stick to the trowel in use, but easily separates from the iron. Whilst this marble coat is drying, another thin coat is to be applied: this is to be well worked and rubbed, and then still another, finer than the last. Thus, with three sand coats, and the same number of marble-dust coats, the wall will be made solid, and not liable to cracks or other defects.
>
> VII, iii, 5–6

The number of coats of plaster he recommends seems like overkill. Before the advent of modern plasters and plaster boards, a traditional specification would have been: apply a 'rendering' or 'scratch coat' of 'coarse stuff' (it was scratched or scored to provide a key to ensure the next coat adhered), a second or 'floating' coat of finer 'gauged stuff', and finally a thin 'skim' or 'setting coat' of lime putty with, perhaps, a little fine sand. There was a 'fancy setting stuff' which contained marble dust. This seems to be what Vitruvius

Plaster Stucco & Fresco

intended to create, since his expression 'does not stick to the trowel' is very similar to his specification elsewhere for lime putty (Chapter 8).

Nowadays flat ceilings are usually of plaster board, and curved ones are supported by expanded metal. Fifty years ago, plastered ceilings were supported by laths, which were split (not sawn) strips of wood – either oak or pine nailed to supporting timbers; occasionally wire netting was used for curved ceilings. At Oplontis, flat, plastered ceilings supported on reeds survived the eruption of AD 79, and Vitruvius (VII, iii, 1–2) describes making vaults:

> Parallel ribs are set up, not more than two feet apart: those of cypress are preferable, because fir is soon spoiled by decay and age. These ribs being set out to the shape of the curve, they are fixed to the ties of the flooring or roof, as the case may require, with iron nails. The ties should be of wood which will not rot or age, such as box, juniper, olive, heart of oak, cypress, and the like, common oak always excepted, which, from its liability to warp, causes cracks in the work.
>
> When the ribs are fixed, Greek reeds, previously bruised, are tied to them, in the required form, with cords made of the Spanish broom. On the upper side of the arch a composition of lime and sand is to be laid, so that if any water falls from the floor above or from the roof, it may not penetrate. If there be no supply of Greek reeds, the common slender marsh-reeds may be substituted, tied together with string in bundles of appropriate length, but of equal thickness, taking care that the distance from one knot to another be not more than two feet. These are bound with cord to the ribs, as above directed, and made fast with wooden pins.

Where plaster was to be applied to wattle-and-daub, he recommends that, in order to avoid cracking of the plaster, two layers of reeds should first be nailed to the walls at right angles to one another; the equivalent of old-fashioned 'cross lathing'.

Unfortunately the word *stucco* is misused. In translation from the Latin it is frequently used instead of 'plaster' and nowadays it is applied to unattractive rendering on nineteenth-century houses. The word is sixteenth-century Italian, apparently from a German word meaning 'crust', and was originally used to mean plaster used to produce architectural detail, such as column fluting, cornices and mouldings, and to make surfaces decorated in relief (Fig. 28). Translators of Vitruvius tend to use it in the modern way to mean any fine plaster finish, including those using gypsum plaster (plaster of Paris).

This fine, quick-setting plaster was known at least from early dynastic Egypt. Gypsum is a naturally occurring mineral, hydrated calcium sulphate ($CaSO_4.2 H_2O$). At a temperature of 130–170°C – a very low temperature compared to the calcining of limestone – it loses water and becomes a 'hemihydrate' ($CaSO_4.½ H_2O$), which rehydrates on the addition of water. However, although gypsum plaster *was* used by the Romans to produce moulded objects, Vitruvius warns, 'A minimum of gypsum should be used, as it sets too quickly and does not allow the work to dry consistently.' In particular, gypsum plaster actually expands as it sets, whereas the lime contracts as it dries.

The first task of the stucco worker, as with the mosaicist, was to mark the outline of the overall pattern on the plaster. Sometimes this was done on the undercoat. Repetitive relief, such as egg-and-dart, could have been made by pressing stamps of wood or terracotta onto the soft plaster. To create relief moulding, stucco was applied on the damp background, sometimes the undercoat, more usually the finish coat. It was then modelled using suitable trowels or special tools (modern 'plasters' small tools') (Fig. 27).

Paint was applied, either suspended in water directly to the fresh plaster (*fresco*), or mixed with a suitable adhesive such as egg white (Pliny) or honey (Vitruvius) and applied to the dry plaster (*fresco secco*). Pigments were mainly mineral. Many of these occurred naturally and some were manufactured (Plate 8).

The unfinished wall paintings in the Casa dei Pittori al Lavoro ('The House of the Painters at Work', IX, 12, 9) at Pompeii gives an idea of how the work was organised. First the large area of colour was applied *a fresco* in three stips, working from top to bottom. The panels were then decorated *fresco secco* with guidelines as underdrawing for ornamental scenes. The main pictorial panels were painted by the master painter, who began by creating preparatory drawings in yellow ochre, over which the final colours were applied. The finished work was polished using wax.

Plaster Stucco & Fresco

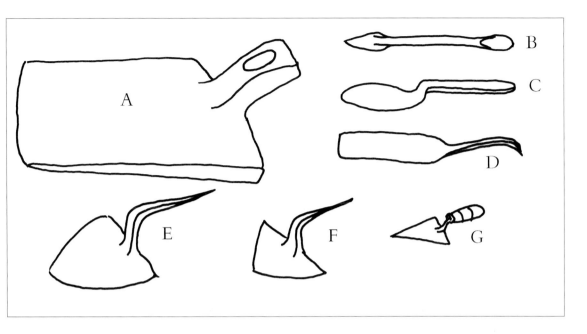

Above: Fig. 27 Plasterers' and stucco workers' tools. A: larry (mixing hoe) from Pompeii 'small tools'; B: from Newstead; C & D: from Pompeii; E–G: trowels in the Museum of London.

Right: Fig. 28 Stucco from the ceiling of the Villa Farnesina, Rome. Augustan Period (late first century BC), now in the National Museum.

Red pigments included cinnabar (mercury sulphide from Spain), which Vitruvius mistakenly calls *minium*. True *minium* (red lead) was made by heating litharge, which was made by made by roasting galena (Chapter 18). Vitruvius calls the product 'second-rate minium'. Oxides of iron are found naturally as red ochre. Realgar, a sulphide of arsenic, is mentioned by Vitruvius and Pliny.

The blue pigment, 'Egyptian blue', was prepared as a frit (glass) by heating together copper, natron and sand, all finely ground. The resulting blue frit was coarsely ground; fine grinding made it lose its colour. Copper carbonate, azurite, occurs naturally. Lapis Lazuli was probably too expensive to be used as a pigment.

Terre verte, a natural green earth of variable composition, was used. Other green pigments included malachite, a natural basic copper carbonate, and verdegris, basic copper acetate, made by leaving copper sheets and vinegar in earthenware pots.

Orpiment, another sulphide of arsenic from Syria and Pontus, was a source of yellow. Among the earth colours called (in modern times) 'ochres', yellow ochre was a clay mineral containing hydrated iron oxide, limonite.

Brown ochre, synopsis, is an earth colour containing alumina and hydrated iron oxide. Brown umber also contains manganese dioxide.

Hydrated lime from oyster shells or marble was a source of white. Lead white (lead carbonate) was produced in a way similar to copper carbonate, by exposing lead to the fumes of acetic acid. This process is important today.

Black (lamp black) was made by burning resinous woods and condensing the smoke on a cold surface. Vitruvius' description of the process might give us an insight into the evolution of the *laconicum* (Chapter 21).

Organic dyes were sometimes used. Blue dyes of vegetable origin – woad, madder and indigo – were used, and Pliny included among pigments Tyrian purple, which was extracted from sea snails (*murex*).

11

BRICKS

Clay is the adhesive plastic component which is usually essential for building with 'mud' as described in Chapter 7. It is a naturally occurring substance found in sedimentary deposits and mainly derived by weathering from geologically older rocks. Superficial deposits of clay are extensively used, mixed with other minerals, such as sand ('brickearth'), which produces red brick, or chalk ('malm'), which produces 'white' (cream-coloured) bricks.

When wet the clay component becomes plastic and can be shaped by moulding, and on drying it shrinks and becomes solid but retains its shape. On wetting, it again becomes plastic. The discovery, at the beginning of the Neolithic period, that the hardening becomes irreversible when the clay is 'fired' at even quite low temperatures led to the invention of pottery. At higher temperatures, depending on the composition of the clay, it 'matures' and becomes much harder. At 1,000–1,080°C in a kiln, which allows plenty of oxygen in the air to reach the charge (in an *oxidising* atmosphere), any iron in the clay is turned flower-pot red; the product is *terracotta* (Fig. 29). At very high temperatures clay fuses and becomes glassy (vitrified). Where air is not allowed in the kiln (in a *'reducing* atmosphere') the iron in the clay fires black at low temperatures but blue at higher ones, a fact which, it is thought, was exploited by mosaic workers. It is common to find pieces of Roman tile that are red on two faces with a blue interior ('liquorice allsorts'). These are caused by a prolonged period of reducing atmosphere, when the combustion in the kiln removes the oxygen from the air and is followed by a brief oxidising one, after the fuel was either spent or deliberately removed so that air was allowed to pass through the kiln.

The building components made of clay that were used by the Romans in masonry were all thin compared to their other dimensions and in practice

Roman Building Techniques

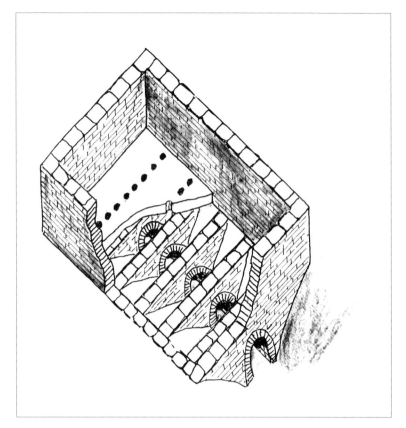

Fig. 29 A typical Roman tile kiln.

everybody calls them 'tiles'. Latin dictionaries, however, translate Vitruvius' word for them, *lateres*, as 'bricks'. He (II, iii) tells us of three sizes of *lateres*: 'the ones we use', which are 1 foot by 1½ feet (*Lydian*), 'those used for public buildings', which are five palms square (*pentedoron*), and 'those of which private houses are built', which are four palms square (*tetradoron*). These, however, are Greek names and Greek bricks. The length of a palm is often given as a ¼ foot (7.4 centimetres). Vitruvius use of the word *lateres* normally implies bricks made of mud, although he does distinguish *later crudus* (uncooked) from *later coctus* (cooked).

By the first century AD most Roman *lateres* were of terracotta, and the sizes were standardised. The dimensions were: *bessales* (8 inches, 19.7 centimetres) square; *sesquipedales* (1½ feet, 44.4 centimetres) square; *bipedales* (2 feet, 59.2 centimetres) square (Fig. 30). Although there was some variation, these sizes were usually adhered to, and buildings tended to be modular, the basic

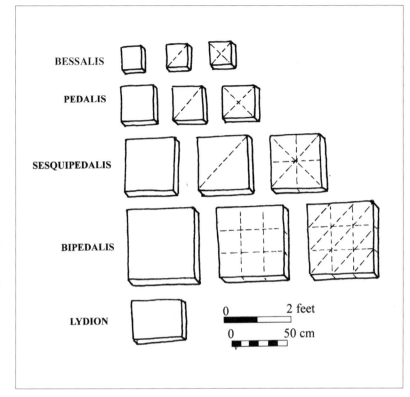

Fig. 30 The various standard sizes of Roman building tiles or bricks. The dotted lines show where they could be cut to make the facing for walls in *opus testaceum*.

measurement being the foot (*pes* = *c*. 296 millimetres). A fired version of the *lydion* was also produced (Fig. 30). Numerous shaped bricks were produced, for example round or segmental ones for constructing columns, and tapered voussoir ones for arches. Vitruvius uses the word from which we derive our word 'tile' – *tegula* (a small covering) – for several terracotta components.

Lateres were used in masonry at every level of construction: wall, reveals (window and door framing), arches, lintels, floors vaults and hypocausts. Other kinds of tile are frequently found to have been used in this way. Roman builders were adept at improvising and recycling.

Recently archaeologists have adopted a horrid 'acronym', 'CBM', for ceramic building material including fired building tiles, roof tiles, pipes and many other terracotta objects except pottery.

12

MASONRY WALLS

Except for foundation trenches dug in solid ground, shuttering was needed when walls were built of *opus caementicium*. Above ground the Roman builders used an integral form of shuttering made of closely fitting stones or pieces of tile, which formed a facing that could be decorative. Vitruvius distinguishes two styles of facing: *opus incertum and opus reticulatum.* In *opus incertum*, which appeared in the second century BC, the walls were faced with small irregular stones. By the first century AD *opus reticulatum* developed. It made use of facing stones, which were cut square on the face side, but tapered into the walls. It was often used decoratively (Fig. 31).

Although *lateres cocti* were being incorporated into structures at Pompeii a century before Vitruvius, his writing suggests that much of residential Rome at least was still constructed with sun-dried mud brick. 'Not every place can have a wall of burnt brick,' he says. Early imperial domestic buildings in Rome were mainly tall apartment buildings of doubtful permanence. In fact, the population of Rome was growing faster than the current methods of building construction could cope. There are many references to collapsing apartments, jerry-built 'skyscrapers', weak foundations and cracking walls. Seneca mentions 'incredible feats that shore up apartment blocks which are already cracking at street level'. The technical problem of attempting high-rise construction in mud brick was compounded by the local by-laws, which, on account of the narrowness of the streets,

> Forbid a greater thickness than one foot and a half to be given to walls that abut on a public way, and the other walls, to prevent loss of room, are not built thicker. Now brick walls, unless of the thickness of two or three tiles, at all

Masonry Walls

Fig. 31 Concrete walls. Top: *opus reticulatum*. Bottom: *opus testaceum*.

events of at least one foot and a half, are not fit to carry more than one floor, so that from the great population of the city innumerable houses would be required. Since, therefore, the area it occupies would not in such case contain the number to be accommodate, it became absolutely necessary to gain in height that which could not be obtained on the plan. Thus by means of stone piers or walls of burnt tiles or rubble masonry, which were tied together by the timbers of the several floors, they obtained in the upper story excellent dining rooms. The Roman people by thus multiplying the number of stories in their houses are commodiously housed.

Vitruvius, II, viii, 17

Roman Building Techniques

Vitruvius uses the word *caementicium*, here translated as 'rubble masonry', suggesting that 'concrete' was being used. However, a hundred years later the satirist Juvenal wrote:

> Here we inhabit a city supported for the most part by slender props, for that is how the bailiff patches up the cracks in the old wall, telling the inhabitants to sleep easily beneath a roof which is about to tumble about their ears.

> Satire 3

The terracotta brick/tile industry burgeoned in the mid-first century, and *lateres* were used for quoins (external corners) and reveals (sides of openings), incorporated in other facings, and, from the mid-first century AD, became used as the principal facing of *opus testaceum*, which was *opus caementicium* faced with fired *lateres* cut into triangles, their apices in the concrete (Fig. 31). These triangles were made by cutting rectangular tiles, either by sawing, or by notching and breaking. This apparently perverse way of doing things is usually, but not convincingly, explained by suggesting that the cut faces of the tiles would adhere better to the core of the wall. What appear to be Roman brick buildings, such as the House of Diana at Ostia, are really of concrete (Plate 9).

Rubble walls of irregular stones, such as flint, often incorporate courses of tiles at more-or-less regular intervals. They provide longitudinal and transverse strength, particularly necessary before the mortar has dried. These are often called 'bonding courses' by archaeologists, although they are anything but bonding, since they actually provide a plane of weakness against toppling through the walls (Chapter 25), and the proper term, used by builders, is 'lacing courses'. The decorative effect is a consideration, and lines of tiles became a feature of Byzantine architecture. It seems reasonable to assume that they marked a stage in the work at which to 'straighten up', and ensured that the work was kept in horizontal courses and constant width. By distributing the pressure as work continued, they would prevent the work, which had not yet become sufficiently solid, from spreading. Another suggestion is that they might originally have been intended to tie ashlar facing onto a rubble core. However, they would not then have needed to go right through the wall.

Masonry Walls

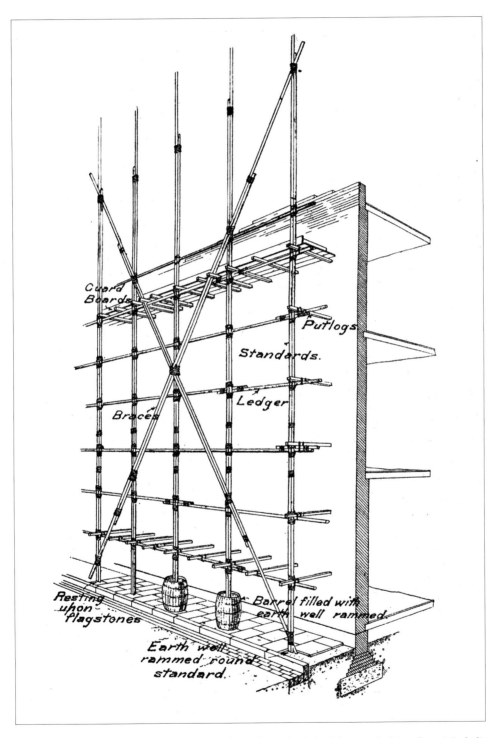

Fig. 32 Bricklayers' scaffolding, consisting of wooden poles joined by rope lashing, from Mitchel's *Building Construction*, 1910.

Roman Building Techniques

Fig. 33 Roman builders at work. The man at the bottom right appears to be mixing lime mortar with a 'larry'. From the tomb of Trebius Justus, Rome.

The 'lacing courses' might possibly be connected with the needs of the builders to obtain access to the work. Scaffolding is a familiar sight on modern building works, nowadays consisting mainly of steel tubing connected by clips that bolt together. Less than 100 years ago, however, these temporary structures were of wood, lashed together with rope. A pre-Second World War book tells the scaffolder how to do the work and, in an illustration that might come from a book for scouts or mariners, shows no less then twenty-nine details of knots and lashings (Fig. 32).

Apart from the odd picture (by an artist, not by a builder) (Fig. 33) we have no details of Roman scaffolding. A clue is sometimes provided by the provision of holes in the masonry to accommodate the short (120–150

Masonry Walls

centimetres), horizontal timbers ('putlogs'), which either socketed into the wall for scaffolding on one side or passed right through it for scaffolding on both sides (Plate 10). These are at a reasonable space horizontally to support scaffold boards and a similar one vertically to allow the builder to reach. They indicate work where the wall was later clad, for example with marble. Their absence might indicate either the scaffolding was free-standing, or that the holes were filled as the work progressed.

13

ROOFS

'Do you know what you've done?' he says. I says, 'No, Sir.' He says, 'You've built a try-angle.' I says 'Have I? I didn't know.' He says, 'That's indestructible, that is. If anybody ever tries to catch you out, just you tell them that any two sides of a try-angle is together greater than the third.'

Bernard Miles

The simplest form of primitive roof was probably thatched and conical, like the round-houses of the Iron Age Britons. Most roofs sloped because, except in desert areas, such as North Africa, their function was mainly to keep the rain out while at the same time shedding the water.

Because weight and weakness of stone in flexion ruled out its use, the structural framework of pitched roofs was of wood. Their main structure consisted of sloping timbers (rafters) from the ridge to the eaves, which would push outwards on the walls if they were not joined across by a 'tie beam' making an isosceles triangle. The outward thrust of the roof is taken up by the beam, which is put in tension. Such a structure puts the load of the roof vertically down the walls; a design factor apparently unknown to the Classical Greeks! In any case, it could not be constructed in stone (Chapter 24).

Because Vitruvius is himself the source of the technical terms he uses, it is difficult to interpret his description of how to make a roof. Figure 34 is an attempt to interpret what he intended to say. With large roofs, vertical posts standing on the beam provided support to prevent excessive sagging of the ridge piece (roof tree) and struts supported the rafters. Vitruvius, however, seems to think that these internal timbers were there merely to fill up the space!

Roofs

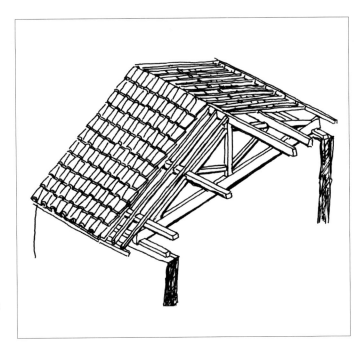

Fig. 34 A possible interpretation of the construction of a Roman roof described by Vitruvius.

Large roofs, with large spans that needed long beams, were limited in length at source: trees. For tie beams lengths of timber can be connected end-to-end by scarf joints (Fig. 16), but there is clearly a limit, especially in timbers which suffer bending loads. If the Romans wanted a bigger basilica, they had to add more aisles with more triangular roof trusses.

Roofing tiles, because of their characteristic colour, texture and, above all shape, are often the first evidence of the existence of a villa site. Terracotta roofing tiles (*tegulae*; singular *tegula*) were used by the Greeks at Paestum as early as the sixth century BC, and Vitruvius implies (II, viii, 19) that they were readily available in his time but, in Rome at least, it seems that fired 'bricks' were not (see Chapter 12).

Tegulae were heavy flat tiles with raised sides and were laid side by side overlapping only *down* the roof. The cracks between them were covered by curved tiles (*imbrices*; singular *imbrex*). The joint was usually filled ('torched') with mortar (Fig. 35).

It sounds simple, but the *tegula* had to be of a surprisingly complex shape (Fig. 36). The ends of the *imbrices* were, in some prestige projects, covered by *antefixes* (Fig. 37).

Roman Building Techniques

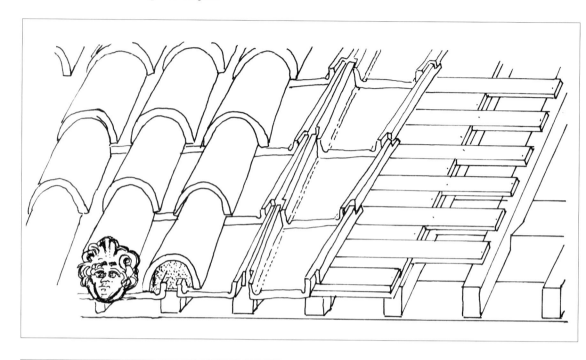

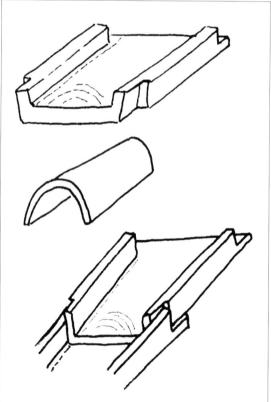

Above: Fig. 35 The construction of a Roman *tegula*-and-*imbrex* roof. Modern examples show that close boarding was not always used under the tiles. Note the mortar torching under the *imbrice* and the antefix on the left.

Left: Fig. 36 *Tegula*-and-*imbrex* roof covering showing the complex shape of the *tegula*.

Below: Fig. 37 Antefixes; moulded from clay and fired, they cover the end of the *imbrices*.

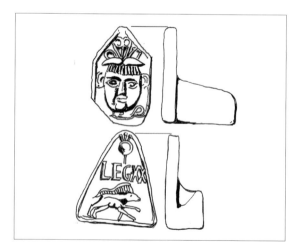

Roofs

In the tile museum at Bridgewater in Somerset are moulds in which, until recently, local tiles were made. They give a good idea of how the Romans might have done it. Tiles were made by 'batting' the clay into a shaped mould, which had been sanded to prevent adhesion and cut with a tensioned wire (Fig. 38). To allow the tiles to overlap, they tapered, and some of the clay had to be trimmed from the lower bottom corners and from the flanges. Once the process was completed, the *tegulae* were usually marked by a freehand impression (usually a rough semicircle). There have been suggestions for this practice. It might have been a sort of signature, but in experimental reconstructions it was found useful to indicate that the fettling process was complete, and also to show which way round the tiles were when a roof was being covered. Sometimes an odd tile was marked with a stamp to indicate where it was made. When tiles were finished, they were allowed to dry in the open air. To judge by the number of footprints of humans and animals (both domestic and wild) many of them were left on the ground. They were then fired in a kiln similar to Fig. 29.

Whereas the sizes of flat Roman building tiles (*lateres*) seem to have been, of necessity, fairly standardised, there are considerable variations in the size of roof tiles from place to place. Their thickness varies as well. If you pick up a *tegula*, you may be surprised at its weight and, including mortar torching, the roof covering of a villa weighed about 90 kilograms per square metre. Actually this is not greatly different from that of many roofs still in use today. Modern clay tiles weigh about 90 kilograms per square metre; pan tiles 60 kilograms per square metre; slates 30 kilograms per square metre. Tiles were probably supported as they are in modern Italy: either directly on the rafters or on boards laid over the rafters.

Many Roman buildings are 'out-of-square', but the error could have been disguised by cutting the *tegulae*, fudging a gap of variable width down the roof, or improvising a wide gusset by substituting a row of inverted *tegulae*. Inverted *tegulae* were also used for decorative effect (Plate 11).

Roman builders also used overlapping 'slates' made from suitably fissile stone from many sources, nailed to battens along the roofs, and tiles are known which more or less imitated these.

A wooden hut on an archaeological training dig was used for years by generations of students – until it became a thing of shreds and patches. A replacement was erected in a more convenient spot and, at an end-of-season

Roman Building Techniques

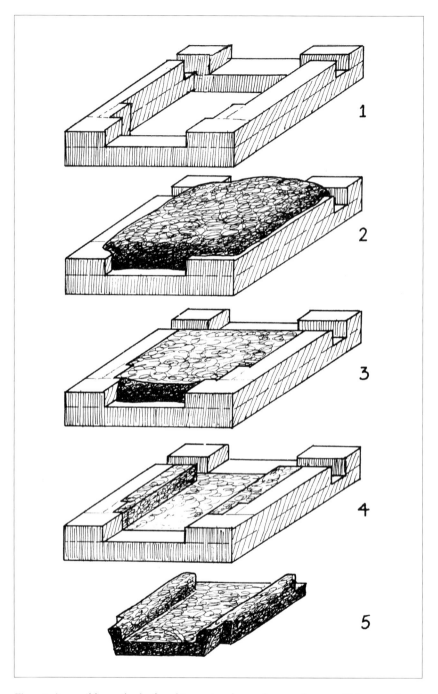

Fig. 38 A possible method of making a *tegula*: 1: the wooden mould is wetted and sanded; 2: plastic clay is 'batted' into the mould; 3: excess clay is cut off the top with a tensioned wire along the mould; 4: clay is cut using the wire across the mould; 5: the tile is demoulded, fettled and set aside to dry. A simpler mould was often used, and the 'cutaways' made freehand.

Roofs

celebration, the old hut became its own funeral pyre. A few seasons later a new generation of students was given the task of excavating the site of the forgotten hut. Would there be anything to prove that it had ever existed? They found a patch of brick rubble which had been put down to fill the puddle outside the door, but the main evidence was negative: a patch about 7 metres by 3 metres where there were no crown beer-bottle tops (today it would be ring pulls).

This story illustrates the problems caused for the archaeologists by the use of biodegradable building materials in the past. Several sorts of organic roof-covering were available to builders in the Roman period. These included shingles, which were thin tiles of split wood, and a wide range of materials of plant origin which are all known as 'thatch', including turf, straw, reed, sedge, heather and other plants.

Shingles were used by the Romans at an early date:

The most suitable shingles are made from hard oak. The next quality comes from other acorn-bearing trees and beech. Shingles are most easily made from resin-yielding trees but these last a shorter time, except for those from pine. Cornelius Nepos reports that Rome was roofed with shingles as late as the Pyrrhic War [which took place 175–280 BC].

Pliny, NH 16

However, Vitruvius says that the house of Romulus on the Capitol, by its thatched roof, clearly demonstrated the simple manners and habits of the ancients.

There are many possible ways in which thatch can be applied. A simple description is that a thick layer is laid along the roof at the eaves. This is held in place, for example, by long hazel sticks ('sways'), which are pegged using sharpened pegs driven into the layer. Rope or flexible plants such as brambles and bark can be used. For temporary thatching of ricks, rope was made by twisting straw with a wimble or twisle – a cranked tool somewhat like a carpenter's brace. Further layers are added, in the same way to the ridge of the roof, where a flexible capping was needed. Straw was suitable, as were sedge and even turf.

Thatching materials are extremely unlikely to survive. They are of interest in that they might affect the actual structure of the roof, which would determine

Roman Building Techniques

the interior space of the building, and any suggested reconstruction, either physical or on paper.

In general a roof is constructed to make rainwater run off as fast as is possible whilst ensuring that the covering does not slide and is not lifted by the wind. *Tegulae*, which were not nailed in place, were probably laid at about 30 degrees and stone 'slates' probably about the same. The recommended angle for modern shingles is 9.5–18.5 degrees.

Since organic roof coverings absorb water, the steepest practical slope is best, not only from considerations of the weight of the roof when wet, but also from those of weathering. Long straw is laid at 45–60 degrees. At an angle of 25 degrees thatching reed lasts up to fifteen years; at 50 degrees it lasts more than forty-five years.

Few roof tiles are found on sites in Roman North Africa, for example at Leptis Magna. It appears that in this area pitched roofs were used only on buildings where this was conventional, as on the baths, basilica, and temples. The flat roofs were probably constructed in the same way as suspended floors (Chapter 14), perhaps using reeds as the supporting layer.

14

FLOORS

The archaeologist comes across many sorts of floors on Roman sites, from scruffy beaten earth, 'puddled' chalk, *opus signinum*, to fine mosaic. Plain wooden floors do not usually survive, but their existence can be deduced, for example, in granaries, where the sockets in the walls must have held joists. In Pompeii and Herculaneum upper floors that were supported in the same way carried concrete floors.

In Book VII Vitruvius gives instructions for laying floors:

> Pavements ... should be executed with the greatest care and attention to their solidity. If the pavement be made on the ground, the soil must be tested, to ascertain that it is solid throughout, then over it is to be spread and levelled a layer of rubble.
>
> VII, i, 1

Here he tells what to do if the concrete floors are suspended on planks, then continues:

> On this is placed a layer of stones, each of which is not to be less than will fill a man's hand. These being laid, the pavement is laid thereon. If this material is new, let three parts of it be mixed with one of lime; but if from old materials, the proportion is five parts to two of lime. It is then laid and brought to a solid consistence with wooden beaters and the repeated blows of a number of men, till its thickness is about three quarters of a foot. Over this is spread the upper layer, composed of three parts of broken tile to one of lime, of a thickness not less than six inches. Over the upper layer the pavement is laid to rule and level, whether tessellated or of opus sectile.

Roman Building Techniques

When laid at the right slope, they are to be rubbed off, so that, if in opus sectile [Plate 12] there may be no projecting edges of the ovals, triangles, squares, or hexagons, but the union of the different joints may be perfectly smooth. If the pavement is composed of tesseræ, the edges of them should be completely smoothed off, or the work cannot be said to be well finished. So, also, the opus spicatum [Fig. 39] of Tiburtine tiles, should be laid with care, that there may be neither hollows on them, nor ridges, but that they be flat, and rubbed to a regular surface. After the rubbing and polishing, marble dust is strewed over it, and over that a coat of lime and sand.

VII, i, 3–4

Tesserae in this context are square pieces of stone or tile, probably cut by breaking them with pincers, as are used today, or possibly on a small iron knife-edged anvil, shaped like a miniature axe (but not as in Fig. 40). They are mainly square; *tesserae* also meant dice. *Opus sectile* consists of relatively thin sheets of coloured stone which, as Vitruvius suggests, are usually cut by

Fig. 39 *Opus spicatum* flooring in the Villa Capo di Bove, Via Appia, Rome.

Floors

Fig. 40 Tomb relief apparently showing mosaic workers making *tesserae*, using scabbling hammers (see Fig. 17). Museum of Roman Civilisation.

lapidary methods into tessellating geometric shapes (*sectila*). They sometimes include pictures, however. *Opus spicatum* consists of small rectangular tiles laid on edge, herring-bone fashion (*spica* is an ear of wheat).

Floor mosaics (the word 'mosaic' is a modern one), made entirely of natural pebbles dating to the seventh century BC, have been found in Asia Minor. In the third century BC in Sicily, mosaic work incorporating both pebbles and cut stone represent the transition to work entirely of *tesserae*. At the beginning of the Imperial period, black-and-white floor mosaics dominate. The pattern and figures were laid in black on a white background. Brightly coloured mosaics gradually became fashionable.

Tesserae were made from coloured stone, often imported from considerable distances. Use was made of the waste products and off-cuts from masons' works. In Britain, red stone was not often available, and use was made of tile – actually this must have been a very convenient and cheap material and commonly British pavements were surrounded by a wide border of large terracotta *tesserae*.

Building and roof tiles sometimes accidentally had blue cores (Chapter 11). At first mosaicists exploited these. Then with suitable choice of clay and firing conditions in the kiln, a range of colours from yellow to black were probably deliberately produced. Glass *tesserae* were sometimes made from broken vessels and, in some instances, were purpose made.

A lot of mosaic work was done *in situ*, but often parts of a pavement, such as pictures or pattern motifs (*emblemata*) were prefabricated. There were two ways in which this might have been done. The *tesserae* were assembled in sand, and cloth or paper was glued to them, or they were glued separately onto a drawing, probably a coloured cartoon. Once they had been set in mortar, the paper or cloth was dissolved off. Only then did the pattern become visible. On a number of pavements one or more of the *emblemata* are set askew for this reason.

Clearly there must have been something like pattern books from which the client could choose. The similarities between mosaics in various regions in Britain lead us to attribute them to individual local workshops (*officinae*) (Fig. 41).

Fig. 41 Suggested areas served by the main mosaic workshops in Britain in the fourth century: Du: Durnovarian; C: Corinian; Db: Durobrivan; P: Petuarian.

Floors

Fig. 42 Coloured mosaic fountain, in the House of the Large Fountain Pompeii (VI, 8, 22).

The word 'grotto' conjures up an image of an artificial cave incrusted with cats'-brains tufa, shells, glass, pebbles and so forth. The eighteenth-century gentleman's conceit was inspired by Roman prototypes. Originally these were natural caves, *nymphaea*, with springs said to be the homes of nymphs, but by the first century BC they became decorative features, often in the apsidal undercroft of rich villas. They were decorated with shells and pumice. Then stucco and marble chips were added, and elaborate patterns were introduced. Wall and vault mosaics used brightly coloured glass *tesserae*, quite independently of floor mosaics, which remained mostly black and white. At Pompeii and Herculaneum there are coloured wall mosaics and highly decorated fountains, which were the descendants of the early *nymphaea* (Fig. 42).

15

STAIRS

Among the problems which face the archaeologist in Roman Britain is that most of the buildings on sites he works on have been demolished and even ploughed flat, so that he is faced with what is, in effect, a multi-period plan. It is usually possible to deduce the different periods (if not the dates) when the various wall footings were constructed, but any attempt at reconstruction in three dimensions has to be based on informed guesswork, since we cannot know how tall the building was.

It has been generally assumed, except in cases where there was a cellar, that the Romano-British villa was single-storied. At Lockleys, Welwyn, there is possible evidence of a suspended concrete floor over a sunken room. If we are to propose an upper floor for a building, we must look for a ramp or evidence of stairs. Attempts have been made to deduce the existence of wooden stairs by examining the plan of a villa. An examination of the structure of wooden stairs and its possible evolution will show the futility of this.

The simplest way of getting to an upper floor is by means of a ladder, which consists of two long timbers (side rails) supporting wooden cross members (rungs) up which the user climbs. To make this easier, broader side rails are erected at a shallow angle, and flat steps (or 'treads') can replace the rungs, giving a 'step ladder'. Either sort of ladder can be erected anywhere in a building to give access through a hole in an upper floor and the discovery of a gap between floor joists is sometimes the only evidence of earlier access to an upper floor in a modernised surviving timber-framed building.

The step ladder becomes stairs (or a staircase) when it becomes a more-or-less permanent feature of a building. The side rails are then called 'strings' and the steps are called 'treads'. The only surviving examples of this are at

Stairs

Fig. 43 Stairs at a shop in Herculaneum (Ins IV, 20) with modern covering against the weather. Beginning with solid stone steps, they continue with solid wooden ones. Above this the slot in the wall suggests a stringer.

Herculaneum, although it can be deduced that they existed in Pompeii. The strings were often supported at the bottom by a solid base of masonry blocks (Fig. 43).

Stairs in Roman buildings could be made of solid baulks of timber supported on strings; or of stone, or concrete, and one of the joys of visiting Ostia is going upstairs (Plate 13) to look over the town.

16

WINDOWS

Many windows were simply openings, perhaps protected by iron grills (Fig. 44). Perforated iron, star-shaped crosses found in excavations might have been riveted at the intersection of the bars of such a grill (Fig. 45) or to a wooden framework. It is possible that, in some instances, they were used to hold panes of glass in place.

Translucent minerals, such as thin marble, selenite and mica, as well as waxed or oiled fabrics, were used to exclude draughts. Glass (*vitrum*) was made by fusing together sand (silica) and the sodium carbonate (soda), which occurred as 'natron' in Egypt. Limestone or lime (see Chapter 8) was added because it made the glass more durable and insoluble in water. Too much lime caused the glass, which is actually a super-cooled liquid, to crystallise or 'devitrify' (literally de-glass). For making clear-glass vessels, small quantities of, for example, manganese dioxide (glassmakers' soap or pyrolusite – a word from the Greek 'fire washer') were added. The amethyst colour these produced counteracted the green caused by the presence of iron in the sand. For decorative glass, small amounts of other metallic oxides were added to the melt to produce bright colours. Window glass was usually left green.

There is little evidence that any glass was made in Roman Britain, however, and it was probably imported in blocks from the Eastern Mediterranean. It was melted in tank furnaces (Fig. 46). In the medieval period, potash glass was made in England, substituting wood ash (mainly from beech) for soda.

Early window glass might have been made by pouring it out onto a sanded or stone surface, and spreading it with a metal or wooden spatula.

Windows

Fig. 44 Iron grill in a window at Herculaneum.

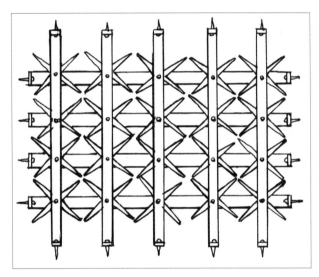

Fig. 45 Iron grill for insertion in a wooden frame.

However, a revolution in glass manufacture occurred at the end of the first century BC with the invention of glass blowing. An iron tube (blowing iron) was inserted into the molten glass (metal), and the gathered material was shaped by rolling it on a flat metal or stone surface (marver). Blowing into the tube produced a bubble. This bubble could be blown into a mould to make a glass vessel, and it has been suggested that the Romans made slab glass by blowing a square bottle shape, which was cut into separate sheets for windows.

To make glass vessels, the bubble was attached to a solid iron rod (puntil), and removed from the blowing iron by making a weak point, perhaps by

Roman Building Techniques

touching it with a drop of cold water, and cracking it off by tapping the blowing iron. It was then reheated and manipulated by shaping, using metal or wooden tools.

Roman window glass was mostly made by the cylinder or muff process (Fig. 47). 'Metal' (molten glass) was collected on the blowing iron and blown into a pear shape. By dipping and re-dipping, more 'metal' was collected until there was a sufficient amount for the intended sheet. After further blowing, the glass was flattened on the marver. After reheating, the resulting bulb was continually rotated, reheated, blown and swung over a pit. The resulting elongated bulb was cut from the iron, and its base was removed. When it was cool enough, the top was also cut off, leaving a cylindrical 'muff', which was cut down the side, opened slightly and put into an oven. It tended to flatten under its own weight, and this is assisted by wooden tools or smoothing blocks. As with all glasswork, the sheets had to be 'annealed', that is cooled slowly, to remove internal stains, which would weaken them.

Fig. 46 A Roman glass tank furnace. (After D. Hill)

Windows

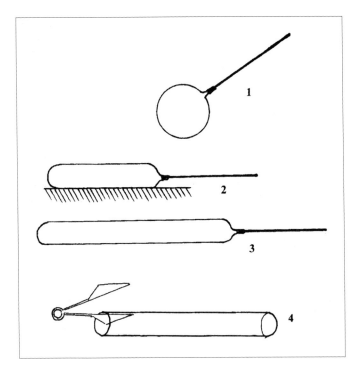

Fig. 47 Sheet glass made by the muff process: 1: a large bulb is blown on the puntil; 2: the bulb is marvered on a flat plate; 3: the resulting 'muff' is stretched by being swung in a pit; 4: the muff is cut lengthwise with shears, before being opened out to form a sheet.

The muff process was still the best way of making flat panes of glass 1,800 years later. It was used to make 293,655 sheets of glass, each 245 centimetres by 1,245 centimetres, in one year, to cover the Crystal Palace, which opened in 1851.

This process made window glass readily available to the rich Roman. Examples with *opus signinum* attached show that window panes were set directly into the masonry. It is likely they were also held in place in wooden frames using lime-and-oil putty, or simply with wooden beading.

The Roman gentleman could even have a conservatory. Martial wrote, around AD 90:

> Your vineyard blooms shut up in transparent glass and the fortunate grape is roofed but not hidden.
>
> *Ep.*, VIII, lxviii

> That your orchard trees from Cilicia may not grow pale and sickly and dread the winter, nor too keen an air nip the tender boughs, glass windows facing

Roman Building Techniques

the wintry south winds admit clear sun and daylight undefiled. But to me is assigned a garret shut in by an ill-fitting window which even Boreas himself could not abide. Is it in such lodging that you cruelly bid your old friend to dwell? As a guest of one of your trees I should be better protected!

Ep., VIII, xiv

17

DOORS

Bronze doors of the Roman period still exist; for example the doors of the Roman senate, now on the Lateran Basilica, and the doors of the Temple of Divus Romulus, in the forum (Plate 14). They clearly copy panelled wooden prototypes, some of which survive at Herculaneum and, as casts, at Pompeii.

On heavy doors the Romans did not use hinges of the types we might consider usual, such as butts or garnets. Their doors were pivoted at top and bottom into lintel and threshold, in which the pivot hole and scuff marks can often be seen. The pivot was a cylindrical bar of metal or purpose made. Folding doors and 'stable' half doors can be seen as plaster casts at Pompeii.

Sometimes doors that opened inwards were fastened by means of a bar across, which dropped into slots on or in the jambs on the inside. At Pompeii there is a cast of a door which was made firm by means of a pole sloping at an angle between the floor and the door. In excavation the common evidence for sliding bolts, which were usually made of hard wood, is the survival of the iron keys. The ingenious mechanism of some locks seems almost to anticipate the invention of the modern tumbler (Yale) lock (Fig. 48).

Tumblers were set in a pattern into the bolt on the inside of the door, and could be pushed up by the key, the fingers of which had to match the pattern. When the bolt was shot, the tumblers fell under gravity, or were pushed by a leaf spring.

The open fronts of shops at Pompeii and Ostia were closed by shutters, which dropped into slots in the stone threshold. The shutters sometimes included a small door, nowadays called a Judas door (Fig. 49).

Roman Building Techniques

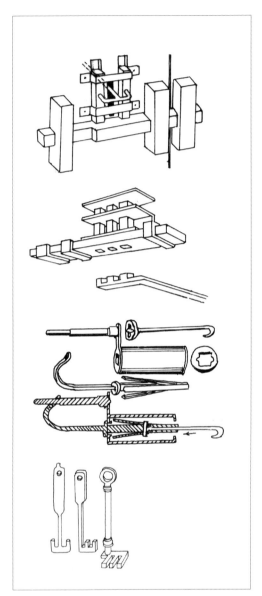

Left: Fig. 48 Roman door locks, padlock and keys.

Above: Fig. 49 Shuttered shop front, Via della Abundanza, Pompeii IX, 7, 10. There is a sally port or Judas door at the right-hand end.

A picture from a villa at Boscoreale, buried by Vesuvius in AD 79, shows the construction of an ornate (and imaginary) door, which includes decorative fittings, apparently including doorknockers (Plate 15).

18

METALS

The smith with his tongs both worketh in the coals and fashioneth it with hammers and worketh it with the strength of his arms.

Isaiah 44:12

The metals which the Romans used in building had mostly been in use for hundreds, if not thousands, of years previously. The processes used for their extraction from their naturally occurring compounds, ores, were crafts, a magic, the results of serendipity, trial and error. We can explain them by modern chemistry but, even today, some of the processes seem akin to cookery.

The common sequence of treatment can be summarised as follows. After mining or quarrying, the ore was crushed and separated from unwanted material (gangue), such as earth and rock, by washing or levigation (most metal ores are heavier, i.e. *denser* than stone). It was then roasted in air. This converts many ores into oxides, and removes some unwanted volatile substances. It is then heated (smelted) with charcoal, which provides a high temperature and removes chemically combined oxygen from the roasted ore, leaving the metal. Commonly a 'flux' is added, to remove unwanted substances as 'slag'. A common flux is limestone, which decomposed and then combined with silica (sand, sandstone etc.) to form slag, which is a glass (Chapter 16) which floats on the top of the metal and can be removed.

MINING → ORE → CRUSHING → WASHING → ORE → ROAST → SMELT WITH → METAL
STONE ⟍ CHARCOAL ⟍
GANGUE & FLUX SLAG

Roman Building Techniques

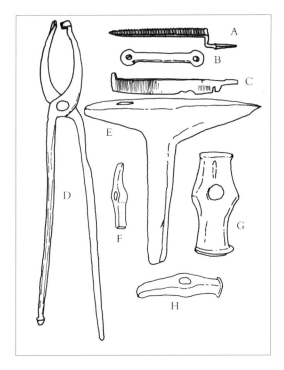

Fig. 50 Blacksmith's tools from Silchester: A: rasp, B: nail header, C: file with notch for setting saws, D: tongs, E: anvil, F: small hammer, G: striking hammer, H: small hammer.

Copper often occurs as a sulphide with iron, in pyrites; the iron was removed as a slag by adding sand at the smelting stage.

Iron was first extracted by the Hittites about 1300 BC. Its ores are common and the metal is ubiquitous on Roman sites. The tools used by a Roman blacksmith are very similar to those in use in recent times. He must have been an essential member of any building team and estate staff (Figs 50 & 51).

Because the high melting point of the metal (1,540°C) was not reached in smelting (which reached about 1,150°C) it did not produce the molten metal, but a soft 'bloom'. Pliny (NH 34) called it a 'spongy mass', which had to be reheated many times and hammered to squeeze out slag, producing 'wrought iron'. It is believed that with a powerful blast of air, as in a modern blast furnace, the Romans did, accidentally, produce liquid iron. This would be 'cast iron', which contains a high proportion of dissolved carbon, and would have been unacceptably brittle for the Romans to use.

The wrought iron contained enough carbon to form the iron carbide cementite. If it is plunged while red hot into cold water ('quenched') it

92

Fig. 51 Blacksmith at work, sharpening tools, Necropolis, Ostia. A wide range of tools and implements is shown. Cast in the Museum of Roman Civilisation.

becomes hardened. Allowed to cool naturally, the cementite disperses, and the metal becomes soft. The hardened iron is brittle, but can be made flexible by reheating it (the temperature being judged by the apparent colour of the thin oxide layer on a polished surface) and quenched again. Quenching was known to the Roman smith; in fact it was known to Homer, who describes Odysseus putting out Polyphemus' eye:

> Just as when a smith plunges into cold water a mighty axe head to temper it – for this is what gives strength to the iron – it hisses violently, thus did his eye sizzle round the olive spike.
>
> Homer, *Odyssey* 9.391

It is believed that Roman smiths sometimes welded (joined by hammering when hot) high carbon iron onto low carbon iron to form the edges of, for example, scythes. The 'mystery' of the smith, which is celebrated in later Saxon literature, included his ability to select different carbon content of iron from the variable content in a bloom, and to judge hardening and tempering

Roman Building Techniques

conditions correctly. Iron can be hardened by hammering while cold ('work hardening') and tools such as axes were also hardened (and at the same time sharpened) by this method.

Copper was the first metal to be extracted from its ores some five millennia before the Roman Empire. Large quantities of it were obtained from Cyprus (hence the Latin *cuprum* for the metal) but deposits were exploited in many places. It is a soft metal but its alloys, particularly with tin (bronze) and zinc (brass) are much harder, and can easily be cast. Some 'bronzes' contain lead. Archaeologists today refer to metal objects, which usually come out of the ground with a matt green patina, as being 'copper alloy'.

Tin is a silvery-white metal. The Romans called it *plumbum candidum* (bright lead). It occurs in Cornwall, and its importance in producing bronze brought the Phoenicians to Britain, the Cassiderides (Tin Islands). Diodorus Sicculus records that tin ingots were taken to an island separated from the mainland at high tide (probably St Michael's Mount), shipped to Gaul, and taken on a thirty-day journey on pack horses to the mouth of the Rhone, to be transported from there by ship. The fact that the Romans were slow in developing the Cornish peninsula suggests that this industry declined in the early Empire because of competition from Spanish deposits.

Lead is a well-known heavy metal with a low melting point. It was called *plumbum nigrum* (black lead) by the Romans. It was used as sheet for example to make boilers, tanks, cisterns, coffins and for pipes (Chapter 19). Alloyed with tin it made solders, which could be used for joining lead sheets and pipes. It was exported from Britain in huge amounts as ingots weighing between 50 and 90 kilograms. Many of these have been found. They are of value to archaeologists because of the inscriptions cast into them. The importance of lead is emphasised by the fact that one of these inscriptions tells us Mendip lead was being extracted by AD 49, within six years of the invasion of Britain.

Zinc was probably not extracted as metal. To make brass, a zinc ore was added to the copper ore before smelting. Some copper ores naturally contain zinc.

19

WATER

Our biological requirement of water – that is, what we need to drink to stay alive – is surprisingly small. In Britain it would be possible to collect this amount by catching rain in a tank about the size of a coffin. This may answer the oft-posed question of how prehistoric people survived in the 'hill forts' on chalk, although the archaeological record, which always seems larded with references to 'eavesdrip gutters' around the houses, does not seem to contain the required storage. The *impluvium* of Pompeiian houses, which predates the building of the aqueduct, was designed to direct the water from the inward-sloping roof of the *atrium* into what looks like a shallow ornamental pool, which in its turn overflowed into a cistern below, with a miniature 'well' to allow water to be raised.

Nowadays, even when we consider water supply in a historic context, we are inclined to think of purity as being paramount. However, the transmission of disease by micro-organisms was only demonstrated 150 years ago. In rural England less than a hundred years ago, water was still taken from ponds and from wells that collected surface drainage, often from sites which were close to cesspits and cemeteries! In towns the rivers carried the sewage and provided water to drink. This is why most people drank beer or, more recently, tea. They knew it was better for them than water; of course, the brewing process had sterilised it. Pliny says 'all water is better for being boiled'.

Wells were commonly used and a variety of linings were employed, especially 'steining' (lining with stone), timber, sometimes with fine carpentry joints, and second-hand barrels. The preservation of waterlogged organic material often makes the excavation of a well rewarding.

Roman Building Techniques

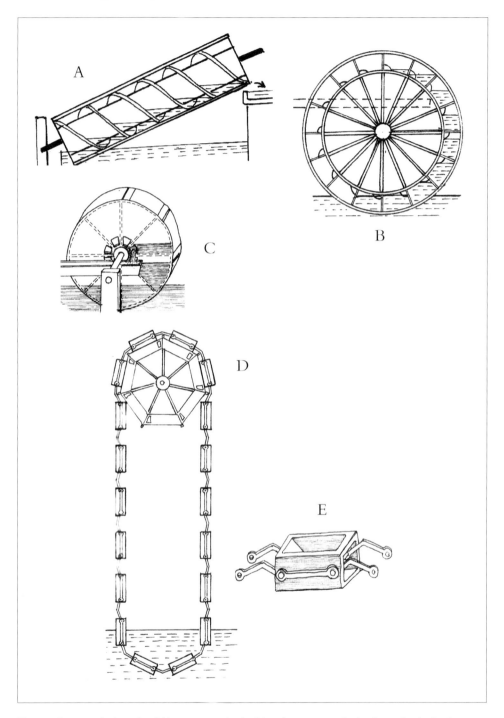

Fig. 52 Roman devices for lifting water: A: Archimedean screw, B: bucket wheel, C: drum or compartment wheel, D: bucket chain, E: detail of one bucket from D.

Above: 1. *Forum Boarium*, Rome. In the foreground is the round so-called Temple of Vesta. In the background is the 'Temple of Fortuna Virilis', probably in fact the Temple of Portunus.

Right: 2. The Pantheon built by Hadrian AD 112, painted by Panini *c.* 1734. (Courtesy National Gallery of Art, Washington)

Above: 3. The *Macellum* at Leptis Magna.

Left: 4. The Arch of Septimius Severus, Leptis Magna, probably constructed for the visit of Severus in AD 209.

5. Castel Sant'Angelo, Rome. The drum is the mausoleum built by Hadrian for himself and his family around AD 135. The bridge, today Ponte Sant'Angelo, was built as *Pons Aelius* by the emperor in 134 as an approach to the mausoleum.

6. The *Pons Fabricius* built 62 BC connecting the *Forum Boarium* to the *Isola Tiberina* in Rome. The span of each of the wide arches is 24.5 metres.

7. Specimens of burnt daub from a Roman site at Welwyn, showing the distinct impressions of the woven wattles.

8. Mural from the Casa del Bracciale D'oro, Pompeii.

9. The House of Diana, a balconied tenement building at Ostia, where commonly the brick facing of the *opus testaceum* was displayed.

10. The Villa of the Quintillii, on the Via Appia. Putlog holes go right through the walls, which must have been covered by decorative stone veneer.

11. Modern tiled roofs near SS Luce e Martina, Rome, showing how the ridge, sloping from right to left, was covered without the use of ridge tiles. Note also the use of inverted *tegulae* down the roof for decoration.

12. *Opus Sectile* of red and green porphyry, gallio antico, Palombino and slate. From the *Domus Tiberiana*, 'Neronian on the Palatine', Palatine Museum.

13. Concrete stairs in a tenement at Ostia. The railings are a modern health-and-safety measure.

14. The bronze doors of the temple of Divus Romulus in the forum, Rome.

15. A *trompe l'oeil* wall painting of a door. From Bocoreale, before AD 79, Naples Museum.

16. Pont du Gard, a famous aqueduct. It is part of a 50-kilometre-long aqueduct that runs between Uzes and Nimes. (Brian Bolton-Knight)

Opposite: 17. A water tower at a crossroads in Pompeii. A fountain fed into the stone tank at the bottom. The stone blocks are a pedestrian crossing which must have acted as traffic calming!

Above: 18. A roadside fountain at Pompeii. The brass tap is, of course, a modern improvement.

19. The charcoal-burning brazier that was still in use in the warm room of the men's section of Forum Baths at Pompeii.

20. The *schola labri* (cold fountain) of the *caldarium* of the men's section of the Forum Baths at Pompeii. In the centre is the *labrum* a cold fountain. At the bottom right a hole has been broken through the face of the wall. It is shown in detail in Fig. 61. 'The labrum should be lighted from above, so that the by-standers may not cast any shadow thereon, and thereby obstruct the light. The schola labri ought to be spacious, so that those who are waiting for their turn may be properly accommodated.' Vitruvius V, x, 4

21. *Tubuli* in the Central Baths at Pompeii, which was being built in AD 79. The tiles on the wall were held in place by iron spikes.

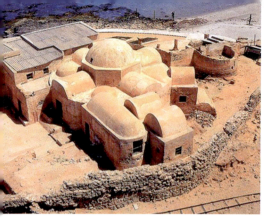

Above: 22. Aerial view of the extraordinary baths of the Villa Silin, Libya. (Photo Libya Soup)

Right: 23. The interior of the church of Santa Maria degli Angli, originally the great central hall of the Baths of Diocletian, converted by Michelangelo in the sixteenth century. It was later somewhat altered by Vanvitelli in 1749. It gives a vivid insight into the appearance of the great Imperial Baths.

Below: 24. The Temple of Aphaia on Aegina. In Greek architecture the slope of the roof was accommodated by a second row of columns.

Right: 25. The 'back gate' at Mycenae, showing the use of a stone lintel.

Below: 26. Temple at Paestum, showing cracked architraves, repaired in modern times.

Above: 27. A true arch, made from unmortared voussoirs. Actually part of the vaulted tunnel through the earth banks around the stadium at Olympia, fifth century BC.

Left: 28. A pendentive containing a mosaic of a seraph in the atrium of St Mark's, Venice (thirteenth century AD).

Water

Water could be lifted by use of a bucket and rope. Windlasses were probably common, although, surprisingly, it is generally agreed that the Romans did not know the crank. The *shadouf*, a device which survived until recently in Egypt was sometimes used. It consisted of a pivoted pole with a weight on one end to balance the weight of the full bucket. In Latin it was called *ciconia* (stork) or *toleno* (lifter).

Vitruvius describes devices for lifting water in Book X, chapter iv. He describes the construction and use of the 'Archimedean' screw pump. It was built of wood waterproofed with pitch (Fig. 52A) and would not be suitable for use in a well, as it had to be set at an angle. We do not know how it was turned. At low inclinations a treadmill would be feasible, but Vitruvius recommends an inclination of 37 degrees; designing a treadmill to work it would be difficult!

Vitruvius describes a 'drum' (*tympanum*), which consisted of a cylinder divided into eight compartments, turning on a horizontal axle. Water entered through slots in the rim, and left via holes near the axle into a collecting trough or 'launder'. It was easier to construct than a screw pump and, since it was mounted horizontally it could have been operated by, or even as, a treadmill. Its obvious limitation is that it could have lifted water to only about two thirds of its diameter (Fig. 52C).

Bucket wheels gave a greater lift than the screw or the drum. In the version described by Vitruvius:

> When it [water] has to be raised higher ... A wheel on an axle shall be made. Round the circumference will be cubical boxes made watertight with pitch and wax so that, when the wheel is turned by treading, carried up full ... will empty into a reservoir.
>
> X, vi, 3

A device clearly related to the *tympanum* is known from the Roman mines in Spain. The buckets are integral with the rim, and empty through their sides. In one mine, used in eight counter-rotating pairs feeding a succession of sumps, water was raised over 29 metres. One of the wheels, incorrectly assembled (the outlet ports should all be on the same side of the wheel) can be seen in the British Museum (Fig. 52B).

Roman Building Techniques

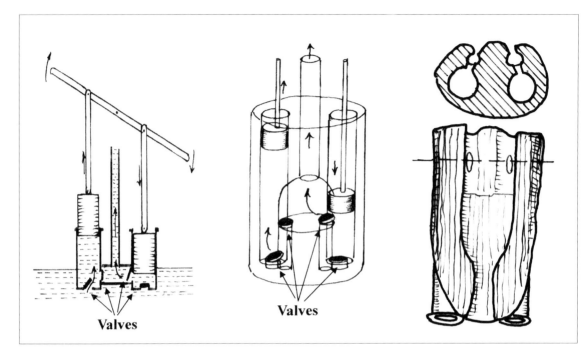

Fig. 53 Roman force pumps. Left: a representation of a double-acting pump. Centre: A diagram to illustrate its mechanism. Right: A sketch of the wooden pump from Silchester, now in Reading Museum.

Another device is described by Vitruvius:

> If a supply of water is needed at a greater height, a double iron chain which will reach the surface is passed round the axle with bronze buckets attached ... The turning of the wheel, winding the chain round the axle, will carry the buckets to the top, and as they pass over the axle they must turn over and deliver what they brought up into the reservoir.
>
> X, vi, 2

A sophisticated example of this device, but using oak boxes rather than bronze ones, was discovered in London recently (Fig. 52D).

Force pumps, especially double-acting ones as described by Ctesibius and Vitruvius, are well known. They were made of wood or bronze. Pistons were moved in two vertical cylinders by a rocker arm. Inlet ducts at the bottom of the cylinders were provided with valves consisting of metal discs, held in

Water

place by 'pins'. The outlet ducts led into a vertical cylinder. The valves were hinged flaps or discs. Since the water was pushed up, not by the pressure of the atmosphere but by the descending piston, the limitations of the height to which this pump could operate would be set by the strength of the plumbing which it supplied (Fig. 53).

Because it was pushed up by atmospheric pressure, the maximum height to which water can be raised by an efficient *lift* pump is about 9 metres. Although this would have been quite enough for many wells – lift pumps were commonly used for this purpose less than a hundred years ago – the device does not appear in books on Roman technology.

Aqueduct means a conduit for water. Cross-country aqueducts, which often brought water from a distant source, were probably not constructed principally from considerations of the purity of the water which they brought, but because of the large quantities of water that were needed by Roman towns and were not locally available. For example, the estimated output of the aqueducts of Rome was about 1,128,000 cubic metres per day, or about 1,100 cubic metres per head of the population. For comparison: the average household consumption in London is about 15 cubic metres per day.

The chief problem confronting Roman water engineers was that they could not economically make or join pipes that were capable of withstanding high pressures. An aqueduct was therefore a covered leet or channel (*specus*) constructed at a shallow gradient. It was not like a modern water main; it could not be turned off. The first aqueducts for Rome – the Appia, 32 kilometres long, constructed in 312 BC, and the Anio Verus, 63 kilometres long, constructed in 272 BC – were, for tactical reasons, underground channels. It is probable that many aqueducts have not been discovered or traced. In many cases the construction of an aqueduct was a prestige project, but the planning, as with more recent railways or canals, must have been the outcome of economic considerations. The source of the water must have been be higher than the point of delivery, and a gradient and size of *specus* had to be chosen that gave a suitable rate of flow. Ideally the route chosen would almost follow the contours, but when the surveyor encountered a hill or a valley in the way of the route he had to consider the relative costs of construction around the obstruction and the more expensive and sophisticated solution such as a tunnel, or one of the spectacular multi-arched structures that are what most people associate with the word 'aqueduct' (Plate 16).

Roman Building Techniques

Sometimes, despite the technical difficulty of sealing the joints, a valley was crossed using what is perversely called an 'inverted siphon', consisting of a lead pipe or earthenware pipes running down the slope and up the other side. In some cases the optimum economic solution was to take the pipe down to a section called a venter (in Greek κοιλια, the belly), supported on a lower, level, arcaded, aqueduct. The siphon needed to begin in a 'header' reservoir, and end in an 'evacuation' reservoir, each of which was supported on a tower that was integral with the arcade supporting the *specus*. Vitruvius does not seem to understand the physics of what is, in effect, a U-tube, and several modern writers seem to have the same difficulty. It is, however, clear that the pressure at the bottom of the 'siphon' would have tended to force apart jointed sections, and might have caused serious problems with seam-jointed lead pipes, which were extremely expensive. The greatest threat to the system, however, would have been where earthenware pipes were used, and they changed direction, especially at a sharp bend, which Vitruvius calls *geniculus* (a knee). On straight sections of pipe the joints tend to be held in place by the rest of the system, but at a bend, particularly a right-angled one, the thrust due to the pressure is applied along the pipe which might open the joint.

If the water must be conveyed cheaply, the following method may be used. Terracotta tubes not less than two inches in thickness are provided, tongued at one end, so that they fit into one another. The joints are coated with a mixture of quick lime and oil, and in the elbows made by the level part of the venter, instead of the pipe, must be placed a block of red stone, which is to be bored, so that the last length of inclined pipe, as well as the first length of the level part may be in it. Then, on the opposite side, where the rise begins, the block of red stone receives the last length of the venter, and the first length of the rising pipe. Thus adjusting the direction of the tubes, both in the descents and risers, the work will never be dislodged.

Vitruvius, VII, vi, 8

The reasons for the specific choice of red (sand)stone are not obvious.

An aqueduct delivered water to a distribution system. Where the flow was likely to vary seasonally, large cistern reservoirs were constructed. Ingenious settling tanks were sometimes provided to intercept sediment. The water was distributed through pipes which could be made of wood, clay or lead, often from a *castellum aquarum* or *castellum divisiorum*. These devices divided the flow either into supplies for different districts or different uses (Figs 54 & 55):

> In the reservoir are three pipes of equal sizes, and so connected that when the water overflows at the extremities, it is discharged into the middle one, for the supply of the fountains, in the second those for the supply of the baths, thus affording a yearly revenue to the people; in the third, those for the supply of private houses. This is to be so managed that the water for public use may never be deficient, for that cannot be diverted if the mains are rightly constructed. I have made this division in order that the rent which is collected from private individuals who are supplied with water, may be applied by collectors to the maintenance of the aqueduct.
>
> Vitruvius, VIII, vi, 2

Fig. 54 Inside the distribution building at Pompeii, where the water from the aqueduct was divided into three.

Roman Building Techniques

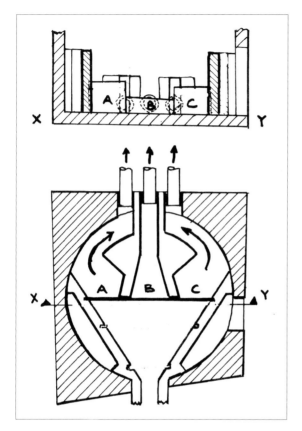

Fig. 55 An interpretation (in Ostia Museum) of the water distribution system at Pompeii (Fig. 54) as rationing water to different consumers, using three sluice gates. A: Domestic, B: Public fountains, C: Baths. This configuration would negate the interpretation of the water towers for local pressure zoning.

Once the water arrived, it was distributed through earthenware pipes jointed with putty, or sometimes wooden ones jointed by iron collars, particularly on level sites. On sloping sites, like Pompeii, lead pipes were probably the best choice. There the town slopes, so the pressure of the water in the distribution pipes would have increased away from the *castellum divisiorum*, threatening the integrity of pipes running to the lower town, which was about 30 metres lower. The water pressure here could have been about 3 atm. The engineers therefore took the supply to water towers; at the top of each was a lead tank (Plates 17 & 18) (*castellum plumbeum*). The function of the towers, like that of the towers at the ends of 'inverted siphons' (see above) seems to have confused modern writers, who don't understand simple hydraulics. For example, J. P. Adam says '*dans laquell, leau perdait sa pression puis repartait dans la distribution urbaine*' (in these the water loses its pressure before returning to the town distribution system). There is no way in which water

102

Water

in a tank can 'lose' its pressure! If the purpose of the towers was to *regulate* the pressure, they provided what a physicist calls a 'constant pressure head', such as that supplying a modern domestic hot water system.

In the header tank of the cold water and central heating systems in a modern house, there is something the Romans did not have: a ball valve, which is a float that shuts off the supply at the right level. The rising water in the header tank would eventually reach the same height (and provide water at the same pressure) as that in the company's water tower or reservoir. If the ball valve fails, there is an overflow pipe, which (hopefully) prevents a

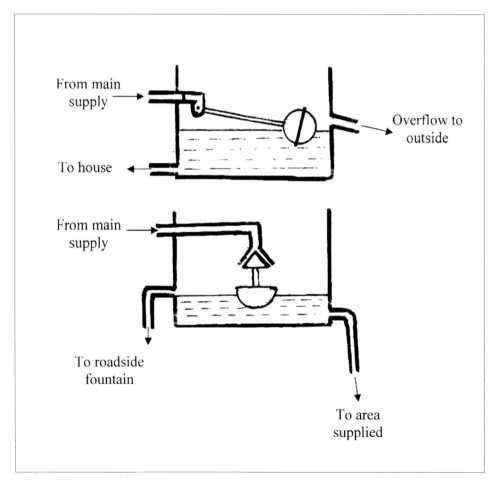

Fig. 56 Constant-level cisterns. Top: a simplified diagram of a modern ball-cock 'water waste preventer'. Bottom: a suggested reconstruction of the reservoir on top of a water tower in Pompeii.

catastrophe in the house, and provides a visible warning that a plumber is required.

Ctesibius introduced a float valve to create a constant head to regulate a water clock (clepsydra). It is possible that at Pompeii, there was a similar automatic valve for the water in the *castellum plumbeum*, but whereas the modern ball cock makes use of a lever crank to increase the force on a washer to cut off the flow, the Pompeian one would leak, providing water for the fountain at the side of each tower. A civil engineer might call the purpose of the system 'pressure zoning' of the areas round the towers (Fig. 56).

20

PIPES

...a supply of hot water gurgles through a labyrinth of lead pipes.

Sidonius Apollinaris to Domitius Letters, Tr. O. M. Dalton (1915)

pp. 34–62; Book II, 2

Terracotta pipes were produced in many sizes and designs and found a variety of uses (Fig. 57D). They were made on a potter's wheel. This limited their length (as with drainpipes and chimneypots in more recent times) to one arm's length (about 600 millimetres). At Pompeii they can be seen typically acting as drainpipes, set in vertical slots in walls. Vitruvius gives a practical tip that if ashes are put in the water when terracotta water pipes are first connected, they will find their way into, and seal, any small leaks.

Lead pipes were usually made from rectangular sheets by rolling them round a cylindrical former, probably of wood, and by soldering along the join. In some instances there was an overlap (Fig. 57C). Somewhat cruder pipes with a sub-triangular cross section resulted when the pipes were rolled without a proper former.

Apparently, to some extent at least, there was a standardisation of pipe sizes. Both Vitruvius and Frontinus, who was *curator aquarum* (water commissioner) for Rome about a hundred years later, produced detailed specifications:

Lead pipes have an area of cross section corresponding to the quantity of water … They should be cast in length of at least ten feet [2.96 metres]. If they are hundreds, they should weigh 1,200 pounds [392.4 kilograms] each length; if eighties, 960 pounds … etc … The pipes get their names from the width of the

Roman Building Techniques

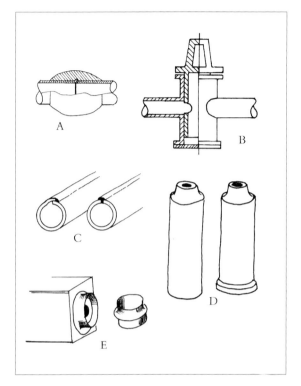

Fig. 57 Plumbing: A: Wiped joint, B: Common design of bronze tap, C: Lead pipes, D: Clay pipes, E: Wooden pipe, showing iron 'collar' which is banged into the end grain.

plates before they are rolled into tubes. Thus when a pipe is made from a plate fifty digits [925 millimetres] in width, it will be called a 'fifty' and so on.

Vitruvius, VIII, vi, 4

These figures enable us to calculate the approximate dimensions of the pipes. Assuming that there was no overlap at the seam, the 'hundred' pipes, for example, were about 590 millimetres in diameter and had a wall thickness of about 6 millimetres.

The Vitruvian standards (and those of Frontinus, below) were intended for regulating the urban water rate, not necessarily for ordinary distribution, and might not have applied on, for example, rural sites.

We can imagine a truly heavy industry for producing pipes in the Roman period. The lead was melted in batches of at least 400 kilograms and poured onto a carefully levelled flat surface at least 3 metres by 2 metres, made of stone or possibly sanded or whitewashed wood with a 6-millimetre raised rim. The resulting sheet was then lifted, trimmed, rolled and soldered.

Pipes

Later, an efflux tube [outlet from the main supply] called a quinaria came into use in the City, to the exclusion of the former measures. This was based neither on the inch, nor on either of the digits, but was introduced, as some think, by Agrippa, or, as others believe, by plumbers at the instance of Vitruvius, the architect. Those who represent Agrippa as its inventor, declare it was so designated because five small ajutages or punctures, so to speak, of the old sort, through which water used to be distributed when the supply was scanty, were now united in one pipe. Those who refer it to Vitruvius and the plumbers, declare that it was so named from the fact that a flat sheet of lead 5 digits wide, made up into a round pipe, forms this ajutage. But this is indefinite, because the plate, when made up into a round shape, will be extended on the exterior surface and contracted on the interior surface. The most probable explanation is that the quinaria received its name from having a diameter of 5/4 of a digit.

Frontinus, *De Aqueductu Urbis Romae*, II, 24

Frontinus lists all the contemporary standard pipes, whether they or not were used. The descriptions become almost impenetrable.

Vitruvius records an old wives' tale that echoes down the ages. Even today one reads some version of the myth that lead pipes were responsible for the fall of the Roman Empire:

Water conducted through earthen pipes is more wholesome than that through lead; indeed that conveyed in lead must be injurious, because from it white lead is obtained, and this is said to be injurious to the human system. Hence, if what is generated from it is pernicious, there can be no doubt that itself cannot be a wholesome body.

Vitruvius, VIII, vi, 10

This is only partially true. Lead is dissolved by soft water (water not containing dissolved 'lime'), which also contains dissolved air. Under these conditions, the lead passes into solution as hydroxide. Lead (like the arsenic beloved of crime writers) is a cumulative poison which does not get passed from the body. Drinking water from lead pipes in a soft-water area can lead to 'plumbism', which causes symptoms that might include: decreased performance in mental tasks, memory loss, paralysis of hand and feet, fatigue, joint pains, anaemia, kidney damage and a bluish-black line at the area where the teeth and gums meet. Although hard water will dissolve lead, in soft-water areas the pipes quickly become coated with lead carbonate (or sulphate). At *Aqua Sulis*, it would therefore have been safe to drink Bath water!

The lead gas and water pipes that were used in recent times before copper pipes with capillary joints replaced them, were not made from sheet, but were extruded. However, the Romans used 'wiped joints' to connect pipes to one another and to other fittings such as taps (Fig. 57B), which were familiar a couple of generations ago (Fig. 57A); the only thing that was missing was the blowlamp, which was invented in the nineteenth century. Books on plumbing do not contain blowlamps before the Primus was invented in 1911. Use was made of plumbers' solder, an alloy of 50 per cent lead with 50 per cent tin. It had a low melting point and was melted in a crucible (or perhaps an iron ladle) and poured onto the joint, where it was wiped into shape using a cloth or, traditionally, a moleskin, lubricated with tallow.

21

HYPOCAUST

For some reason the Romans neglected to overrun the country with fire and
the sword, though they had both of these; in fact, after the Conquest they did
not mingle with the Britons at all but lived a semi-detached life in villas. They
occupied their time ... building Roman roads and having Roman baths. This
was called the Roman Occupation.

Sellar & Yeatman, *1066 & All That*

Roman baths are ubiquitous. Every Roman site, domestic, commercial or
military, seems to have a suite of baths. Some were small and domestic (*balneae*),
like Welwyn Roman Baths, which were preserved by the Welwyn Archaeological
Society in a vault under the A1(M) motorway. Some, like the vast public ones
(*thermae*) are among the most spectacular buildings of the early empire and
the story of their evolution is important both because they made use of almost
every available material and technology, and because they were influential in the
development of the architecture of late antiquity and beyond.

Welwyn Roman Baths provide a good and accessible example of a simple
suite of Roman baths (Fig. 58). They consist of four rooms, three of them
graduated in temperature from unheated to very hot. The fourth was the
furnace room (*praefurnium*) where a slave stoked the wood-burning fire and
maintained the hot water supply from the boiler over it (Fig. 59).

The bathers, accompanied by body slaves, came into the unheated room
(*frigidarium*) and were undressed, putting on wooden-soled sandals before
going into the warm room (*tepidarium*), which was as hot as a summer's day.
They became used to this raised temperature; their bodies were oiled before
they went into the hot room (*caldarium*), where the floor was too hot for bare

Roman Building Techniques

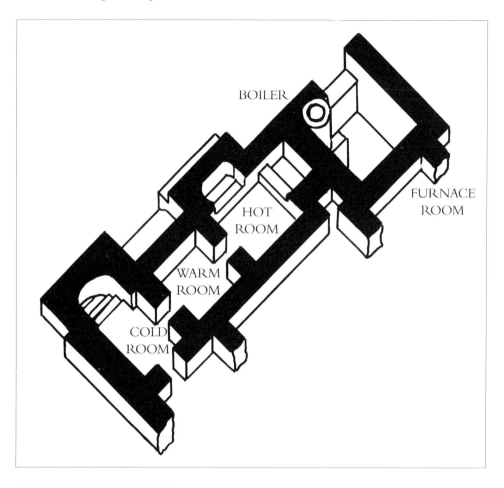

Above: Fig. 58 Simplified plan of Welwyn Roman Baths.

Left: Fig. 59 The boiler of the baths of a villa at Boscoreale. It is made of sheet lead, with a bronze bottom. Note the wiped joints. The branched pipes, with a tap in one branch, allowed hot water from a header tank to be mixed in the right proportions. It is clear that there was no appreciation of convection.

feet (hence the wooden-soled sandals) and the walls were too hot to touch. The humidity and the temperature in this room were very high, (probably 100 per cent RH and about 40°C) and the sweat that was induced couldn't evaporate. The dirt, sweat and oil were scraped from their bodies before they bathed in a hot tub. Afterwards, they went back to the *frigidarium*, where they stepped down into a shallow cold bath and had cold water poured over them. Dried and dressed (we really don't know what everyday dress the well-to-do Romano-Briton wore) they returned to the house. Their bath probably took all afternoon, and, perhaps, every afternoon. The baths were used by both men and women (at the same time) and bathing was a sociable occupation. The baths were a place to meet friends, sing, tell stories, gossip, discuss problems and pass the time of day.

Sweat baths, such as Finnish saunas, Turkish *hammams* and Native American 'medicine sweats', can be found in many cultures all over the world and at every period from the distant past to the present day. Herodotus described Scythian sweat baths, which were made more intoxicating by the use of the vapour of hemp seed in the fifth century BC. Attempts have been made to discover cultural connections between the places where they occur. Sometimes these are convincing, but it is likely that sweat baths have often occurred spontaneously.

In the Hellenistic period, circular sweat baths were heated by a brazier or a simple fire in the centre. This is the hot, dry room or *laconicum* described by Vitruvius in public baths:

> The laconicum [dry sweat room] and sudatoria [steam sweat rooms] are to adjoin the tepid apartment, and their height to the springing of the curve of the hemisphere is to be equal to their width. An opening is left in the middle of the dome [of the laconicum] from which a brazen shield is suspended by chains, capable of being so lowered and raised as to regulate the temperature. It should be circular, that the intensity of the flame and heat may be equally diffused from the centre throughout.
>
> V, x, 5

For the temperature to be adjusted by the suspended shield (*clipeus*) the source of heat must have been a brazier or the like which was inside the room. The other rooms of early baths were heated in this way. The *tepidarium* of the

Roman Building Techniques

forum bath at Pompeii was still being heated by a brazier at the time of the eruption in AD 79 (Plate 19).

The construction of baths was revolutionised by the invention of the hypocaust. The word hypocaust is clearly of Greek origin. '*Hypocauston*' literally means a place heated from below, and although there is some evidence there had been earlier experiments with under-floor heating, the invention appears to have occurred in the first century BC. Several Roman writers, including Valerius Maximus (IX, 1, 1) and Pliny (NH 9, 54, 9) ascribe it, incorrectly, to Sergius Orata, who was active in the last decade of that century. There is plenty of archaeological evidence that this attribution is incorrect, although a real Orata did exist. The stubborn persistence of such spurious traditions is illustrated by the story of (mythical) Adam Thompson, who, despite the best efforts of his inventor, H. L. Mencken, to correct the hoax, has now been repeatedly credited for over a hundred years with having introduced the first bath tub into the United States.

Vitruvius describes the construction of a hypocaust:

> The floors of the hot baths are to be made as follows. First, the bottom is paved with sesquipedales [see Chapter 11] inclining towards the furnace, so that if a ball be thrown into it, it will not remain therein, but roll back to the mouth of the furnace; thus the flame will better spread out under the floor. Upon this, pilae of eight inch bricks are raised, at such a distance from each other, that bipedales may form their covering. The pilae are to be two feet in height, and are to be laid in clay mixed with hair, on which the above-mentioned bipedales are placed, which carry the pavement.
>
> V, x, 2

In practice there was considerable variation of detail. The bottom or sub-floor was made of any firm material: concrete or even compact subsoil was used. There is no theoretical or practical reason for it to slope. However, 'drainage channels' are frequently reported in the sub-floor of hypocausts. They end under the furnace, but we can only speculate what purpose the builders thought they served.

Pilae were often made of *bessales* (see Chapter 11), which were laid with clay or mortar, but were often purpose-made (Fig. 60) or improvised, for

Hypocaust

Fig. 60 Various forms of *pilae* for hypocausts. The bottom two are improvised: one from a *tubulus*, the other from two *imbrices*.

example from *imbrices* or *tubulae*. They were often taller, particularly in later hypocausts, 3 feet (*c.* 91.5 centimetres) being the most common, but it is remarkable that their spacing usually complies with Vitruvius' description, so that the floor structure consists of *bipedales*. As a result there is a degree of standardisation of dimensions; those of even the most unconventionally laid-out bath buildings are likely to be based on a module of 2 feet.

22

CHIMNEYS

Vulcan, drawn by the glowing flue, pants forth his flames and whirls them up through the channelled walls...

Ausonius, *Mosella*

There is one essential detail missing from the Vitruvian description of the hypocaust. There is no mention of rising flues, or, as we might call them, 'chimneys'. These must have been essential, since it is convection, the rising of hot gases, which causes a fire to 'draw' in the confines of the furnace and takes the heat underneath the floors. It is possible to suggest how these flues developed. Their evolution seems to have been rapid, and the following account is merely a plausible hypothesis.

The first chimneys seem to have been simple, cylindrical terracotta pipes (Fig. 57D) rising from the space under the floor. To carry rain or waste water downwards, the narrow neck of each pipe would have pointed down. However, in the heated rooms of the Forum Baths at Herculaneum, probably constructed in the second half of the first century BC, they can be seen inverted, and they carried hot gases up into a 'smoke tunnel' in the thickness of the vault (Fig. 61). These probably represent the earliest 'chimneys'.

A possible second stage of the evolution of chimneys, in the first century BC, can be seen in the hot room of the Forum Baths at Pompeii, where a hole accidentally knocked through the plaster allows us to see that it is separated from the masonry of the wall by a cavity (Plate 20). On close examination, this cavity is found to be bridged by broad terracotta cones, which are the lugs of *tegulae mammatae*: terracotta tiles with projections near each corner (Fig. 62). Sometimes, elsewhere, according to some

Chimneys

Above left: Fig. 61 The hot room of the Forum Baths at Herculaneum, showing the 'smoke tunnel' in the thickness of the vault.

Above right: Fig. 62 A corner beside the apse (*schola labri*) in the caldarium of the Forum Baths at Pompeii. Plaster and tiles have been removed, showing the lugs of the *tegulae mammatae* bridging the gap between the plaster of the *schola* and the wall.

authorities, the lugs were perforated, so that fastening nails could pass through them.

Tegulae mammatae are not mentioned by Vitruvius in connection with hypocausts, but only for the purpose of creating what modern builders call 'dry lining': cavities are made inside walls to prevent the ingress of capillary moisture and provide extra insulation:

> If a wall is liable to continual moisture, another thin wall should be carried up inside it, as far within as the case will admit; and between the two walls a cavity is to be left lower than the level of the floor of the apartment, with openings for air ... If, however, there be not space for another wall, channels should nevertheless be made, and holes from it to the open air. Then bipedales

Roman Building Techniques

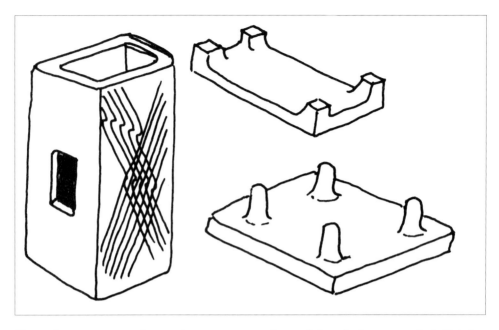

Fig. 63 Some purpose-made ceramic components used as wall flues for hypocausts. Left: *Tubulus*; top right: *Tegula hamata*(?); bottom right: *tegula mammata*. *Tubuli* were sometimes held in place by T-shaped iron spikes banged into the masonry, as in the Central Baths at Pompeii. *Tegulae mammatae* are often illustrated with perforated lugs for nails. At Pompeii the lugs were solid, and the nails were banged through.

are placed on one side, over the side of the channel, and, on the other side, piers of bessales, are built, on which the angles of two tiles may lie, that they may not be more distant than one palm from each other. Over them tegulae mammatae are fixed upright, from the bottom to the top of the wall; and the inner surfaces of these are to be carefully pitched over, that they may resist the moisture; they are, moreover, to have air-holes at bottom, and at top above the vault.

VII, iv, 1–2

The beginning of this reads as if, perhaps unwittingly, Vitruvius was actually describing the structure of a hypocaust! In any case it is not difficult to imagine, since the bathrooms must have been damp, either from condensation or splashing, that someone tried lining the walls, and later had the bright idea of joining the cavity to the hypocaust. This was to have a revolutionary effect on the evolution of architecture. Several kinds of components then evolved as 'chimneys' (Fig. 63).

Chimneys

When a room was lined with *tegulae mammatae*, a problem arose with detailing openings such as doors and windows, and the next stage in the evolution of chimneys was a solution to this which can be seen, for example, at Pompeii in the baths of Julia Felix and in the baths of the House of the Faun. It was the introduction of rectangular terracotta pipes (Figs 64 & 65) to block off the cavity.

Fig. 64 View looking down into the flues in the hot room of the baths of the House of Julia Felix (II, 4, 3-12) (see Fig. 63).

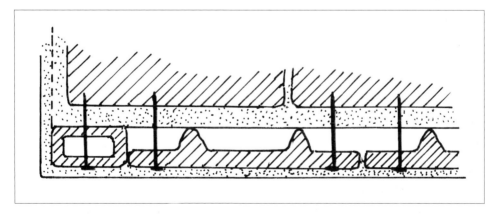

Fig. 65 Plan of the cavity wall of the *caldarium* of the House of Julia Felix at Pompeii. The nails were hammered through the tiles, and it was quite usual for them to be nailed to a wall that had been smoothly rendered.

Roman Building Techniques

Fig. 66 Unused *tubuli* awaiting installation in the Extramural Baths at Herculaneum.

Within a few years, around AD 70, it was realised that a larger version of these pipes, *tubuli*, called the 'box flue tile' by archaeologists, could be used instead of the *mammatae*. They were being installed in the Central Baths at Pompeii, which was under construction on a site that was cleared after the earthquake of AD 62, at the time of the eruption of Vesuvius in AD 79 (Plate 21), and a stack of new, unused, *tubuli* can be seen in a room near the entrance of the Suburban Baths at Herculaneum (Fig. 66), which were being refurbished at the time of the eruption.

It is astonishing that at this time the builders had not invented a sensible way to fix either the shallow or the full-size type of *tubulus*, nor the *tegula mammata*. After the walls of the rooms had been finely rendered, the tiles were put in place, and then nails and clamps were crudely hammered either through the tiles or between them. It seems likely that this practice must have caused a lot of breakage. It gave rise to a more-or-less regular pattern of unsightly pits in the rendering. These would, of course, have been invisible behind the *tegulae* or *tubuli* and the wall plaster when the work was completed.

Many different patterns of *tubuli* were produced, and it became common for tile-makers to comb or to print patterns on them with a roller. These acted as keying and enabled the builders to stick the flue tiles onto the walls with mortar, and then ensured that the plaster adhered to them. Even a small

Chimneys

fragment of a combed or stamped *tubulus* in plough soil is a clear diagnostic clue to a field archaeologist.

Another tile that was used in flues seems the best candidate for the Latin *tegula hamata*, 'hooked tile'. They can be seen in the House of the Labyrinth at Pompeii, where they were used to create a hollow vault in a heated room. Other methods of creating cavity walls were invented or improvised. Clay bobbins of several sizes and shapes were nailed to the walls to support flat tiles or even thin flat stones, as in the fort baths at Bearsden (Fig. 67). Roofing *tegulae* were sometimes nailed, clamped or plastered vertically. These can be seen, for example, in the House of the Surgeon at Pompeii (Ins. 1, 9–10), and the bottom of the stairs at Lullingstone Roman villa.

To make the hot gases go through the hypocaust by convection there had to be outlets at the top of the 'chimneys'. In the Forum Baths at Herculaneum the terracotta pipes opened into the 'smoke tunnels', which opened at one end just like windows (facing across an alley to reduce the chances of wind

Fig. 67 Ceramic bobbin used to hold flat tiles away from a wall in a heated room. Photographed *in situ*.

pressure blowing back down the pipes). Cavity walls, created by using *tegulae mammatae*, bobbins or *hamatae* allowed the builders to create openings to the outside either by making small vents in the wall or through suitably placed pipes in the roof. However, when *tubuli* were introduced the horizontal flow of gases was restricted. It is unfortunate for us that the tops of walls rarely survive to show how the problem was solved. One solution for which evidence does exist is that of the 'exhaust manifold', made of *tubuli*, each with one side wall removed, laid horizontally across the tops of the vertical flues which led to the outlets (Fig. 68). The same arrangement was also used to take the gases around obstructions, such as windows. These survive, for example, under the windows in the Hadrianic baths at Leptis Magna and in the small addition to the Hunting Baths nearby.

Rooms in baths were usually vaulted or domed, and a number of ingenious ways were invented to make hollow barrel vaults. Archaeologists are tempted to suppose these to have been heated because they might have connected with the 'tubulation' of the walls (Fig. 69).

An idea that bedevils archaeology is provided by the Hunting Baths at Leptis Magna. Because this untidy conglomeration of exposed vaults and domes is

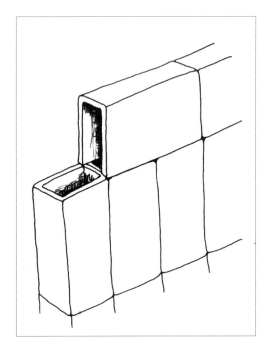

Fig. 68 'Exhaust manifold' or 'collecting channel' at the top of a wall with a lining of *tubulae*, under the window of the Hadruianic Baths at Leptis Magna. In this area *tubuli* were not combed.

Chimneys

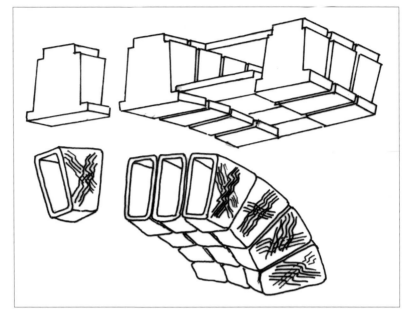

Fig. 69 Hollow vault construction. Top: Suggested reconstruction of the the vault at Chesters. Bottom: Reconstruction using voussoir *tubulae*.

almost the sole surviving published example (there is another, lesser-known, at the villa at Silin, Plate 22) it has been seized with delight by illustrators as representative. All suites of baths, it is implied in 'reconstructions', *must* have looked like this. Several facts have been ignored. One is that Leptis is in a hot, dry desert and is not subject to rain and frosts. Another is that conventional pitched-tile roofs were not usual on domestic buildings in North Africa. Unfortunately, the roofs of Roman baths seldom survive. However, the Great Thermae of Hadrian's villa at Tivoli, admittedly a different scale of building, shows that the vaults (but not the domes) were disguised by conventional roofs, and the discovery of eaves-drip drains (into which a pitched roof shed water) at Chesters has changed the archaeologist's original concept of an exposed vault.

With modifications the Romans adapted the hypocaust heating system for under-floor heating in living and dining rooms, particularly in Northern provinces. For floor heating a small number of vertical flues, usually *tubuli*, were set in the walls to act as chimneys. Floors were supported by *pilae* or piers, or the hot gases were confined to channels. In composite or 'union-jack' hypocausts, hot gases from the furnace went through a main channel into a small, pillared hypocaust in the centre of the room, and were then distributed through smaller channels to the walls.

23

BATHS COME OF AGE

In the first century AD Seneca visited the villa to which Scipio had retired after conquering the Carthaginians in 202 BC.

> I am at the country-house which once belonged to Scipio Africanus himself ... [in] the old-fashioned narrow, dark baths, for our ancestors did not think that one could have a hot bath except in darkness. In these baths there are tiny chinks – you cannot call them windows – cut out of the stone wall in such a way as to admit light without weakening the fortifications.
>
> *Ep. Mor. 86*

It wasn't just fashion that made these baths cramped and gloomy. The design of baths in Scipio's days was dictated by the way they were heated. Early Roman baths used braziers burning charcoal. They could not heat large spaces, and the windows had to be small in order to retain the heat.

About a century after Scipio's day, the invention of the hypocaust provided a degree of heating a little better than the use of braziers, but heat was inefficiently conducted into the rooms through thick concrete floors, and rooms remained narrow and dark. A real change was made possible by the introduction of cavity walls as rising flues.

Seneca writes of technical improvements:

> We know of certain devices which have been invented in living memory, such as windows that let in clear light through transparent panes, suspended [hypocausted] baths with tubuli let into the walls to diffuse the heat while

maintaining an even temperature both at the bottom and the top of their spaces.

Ep. Mor. 90

Much more heat was transferred into the rooms when the hot gases from the hypocaust turned the walls into additional heating elements. The walls of a room usually have a larger surface area than the floor, and the heat passed through a thinner layer of the material than the concrete floors. It was possible to heat larger rooms.

The bathers had to wear wooden-soled shoes (Fig. 70) because the floor was too hot to stand on with bare feet and, as Cornelius Fronto found, the walls could be too hot to touch. He wrote:

While my slaves were carrying me here from the baths as usual, they carelessly dashed me against the blistering entrance, so that my knee was both scraped and burnt…

'To Marcus Aurelius' (*Ad M. Caes.* 44, 59)

We are surprisingly sensitive to radiant heat. If you go into a cold room on a winter's day and turn on an electric fire, the temperature of the room, as measured by a thermometer, is hardly affected, but you *feel* warmed – by the radiant heat. If you are lying on the beach on a summer's day and a cloud goes over the sun, you quite *suddenly* feel cold. Again, the air temperature has not changed; you have been deprived of the sun's infra-red radiation, which travels

Fig. 70 Sole of a wooden bathing sandal from Vindolanda.

Roman Building Techniques

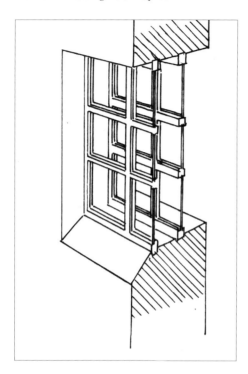

Left: Fig. 71 Roman 'double glazing'. (After Schalles)

Below: Fig. 72 Testudo and boiler: A: The Stabian Baths, Pompeii; B: A villa at Boscoreale. It is unusual to find a testudo in a small suite of baths.

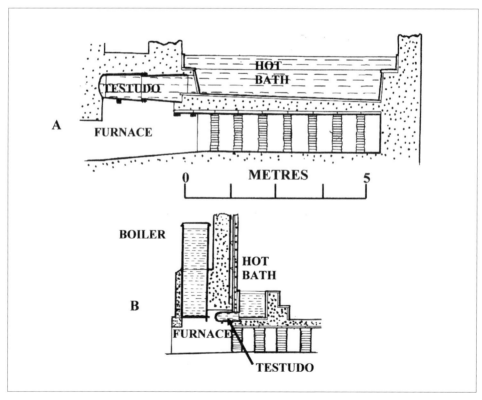

Baths Come of Age

at the speed of light. The effect of the heat radiated from the heated walls of the baths was that some rooms could be provided with large unglazed windows. Even if it were dark outside, the bathers would feel as if the sun were shining.

Seneca contrasts Scipio's baths with the ones common is his own day:

> Who in these days could bear to bathe in this fashion? Nowadays people regard baths as fit only for moths if they have not been so arranged that they receive the sun all day long through the widest of windows, if men cannot bathe and get a tan at the same time, and if they cannot look out from their bath-tubs over stretches of land and sea. The establishments that drew crowds and won admiration when they were first opened are avoided and downgraded into the category of venerable antiques as soon as luxury has invented some new device ... In early days there were few baths and they were not fitted out for display.
>
> *Ep. Mor. 86*

A room that guide-books call *caldarium* in baths, for example in the Thermae of Caracalla (Fig. 73), which has large windows round the south-west side and another in the Forum Baths at Ostia, which clearly had unglazed windows on the south-west side, might more accurately be called *heliocaminus* or *cella solaris*, sun room.

It is possible that, by contrast, the hottest rooms were 'double glazed' by having two frames set one behind another (Fig. 71). This arrangement has been recognised in the Suburban Baths at Herculaneum and the Forum Baths at Ostia.

The temperatures that could be achieved by the hypocaust system were astonishing. Seneca refers to

> ... the heat which men have recently made fashionable, as great as a conflagration. So much so that a slave condemned for some criminal offence ought now to be bathed alive. It seems to me that here is no difference between 'the bath is on fire' and the 'baths are warm'!

In some instances, in order to make the hot tub even hotter, an auxiliary heat exchanger, called testudo (tortoise) of bronze extended it over the furnace (Fig. 72A).

125

24

THERMAE

Wasn't it fun in the bath tonight, with the cold so cold and the hot so hot?
A. A. Milne, 'Vespers', from *When We Were Very Young*

Seneca gives a very vivid idea of the activity in public baths in a letter he wrote in the middle of first century AD.

I would die if silence were as necessary to study as they say. I live just above the bath house. Imagine all these kinds of voices ... While the sporting types take exercise with dumb-bells, either working hard or pretending to do so, I hear groans; every time they release the breath they have been holding, I hear sibilant and jarring respiration. When I meet some idle fellow content with a cheap massage, I hear the smack of a hand on the shoulders, and, according to if it is open or closed when it strikes, it gives a different sound. If a ball-player appears on the scene and begins to count the scores, I'm finished! Suppose there is also some brawler, and a thief caught in the act, and a man who likes the sound of his own voice while taking his bath. Then there are the bathers who leap into the pool, making a mighty splash. But all these people at least have a natural voice. Just imagine the shrill and strident cries of the attendants who pluck the hair from the bathers' bodies, who never cease their noise except when they are plucking the hair from somebody's armpits and making another scream instead of themselves. Then there are various cries of the pastry cooks, the sausage-sellers, and all the hawkers from the cook-shops, who advertise their wares with a sing-song all their own.

Ep. Mor. 56

With the introduction of heated walls there came the possibility of an entirely new class of public building: baths with very large rooms and very large windows (facing south-west because Romans preferred to bathe in the afternoon sun). Unlike the basilica or the temple, they had no conventional structure and their evolution is therefore of outstanding importance in the development of building technology.

Rich patrons, ever anxious to impress the common people with their benefactions, seized upon the idea of making baths into impressive and universal social foci. Baths were created for the populace of cities by emperors. The first were built in Rome between 25 BC and 19 BC by Agrippa; they were later brought up-to-date with heated walls. The creation of massive Imperial *thermae* began with the baths of Titus in AD 81 and eventually in Rome alone there were eleven. Their locations and plans can be deduced from surviving remains and records made of them, particularly by Palladio in the eighteenth century and by Piranesi in the eighteenth. Of particular interest to the modern visitor are the Baths of Diocletian, built in AD 298–306, and part of which was converted by Michael Angelo into the Church of Sta Maria degli Angeli (Fig. 71) and the 'Baths of Caracalla' (Thermae Antoninianae), built AD 206–217.

The massive ruins of the Thermae Antoninianae survived large-scale destruction for building materials in the Middle Ages because of their location outside the city. They demonstrate what is, in effect, the peak of concrete construction. They consist of a central block, which is the baths proper, standing in a garden precinct *c.* 410 metres by 393 metres (about 16 hectares).

The main block was *c.* 225 metres by 115 metres. As well as being overwhelming in its scale, it is a masterpiece of planning, consisting of two intermeshing bathing circuits set out symmetrically on the short axis, along which lie the swimming pool (*natatio*), a vast, three-bay, vaulted *frigidarium*, a smaller *tepidarium* and a circular domed hall usually called the *caldarium*, which was probably a sunbathing room, referred to as '*cella solaris*' in the *Historia Augusta, Caracalla*, 9. This is 35 metres in diameter, with windows beneath which were heated baths. It is easy to imagine the bathers (like those described by Seneca a couple of centuries earlier), bathing and getting a tan. On either side of the precinct were wide exedra, each encircled by a barrel-

Roman Building Techniques

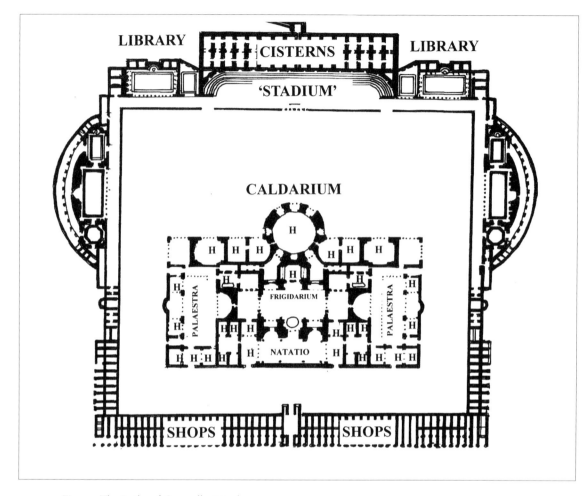

Fig. 73 The Baths of Caracalla. H = hypocaust.

vaulted corridor and containing three rooms. In the corners of the precinct were libraries, and across the south-west side was either a cascade, or possibly a stadium with banks of seats raised on a cistern.

25

STRENGTH & STABILITY

And it came to pass, as they journeyed from the east, they found a plain in the land of Shinar; and they dwelt there. And they said to one another, Go to, let us make bricks, and burn them thoroughly. And they had brick for stone, and slime they had for mortar. And they said build us a city and a tower, whose top may reach unto the heaven; and let us make us a name, lest we be scattered abroad upon the face of the whole earth. And the LORD came down to see the city and the tower, which the children of men had builded.

And the LORD said, Behold the people is one, and they have all one language; and this they have begun to do: and now nothing will be restrained from them, which they have imagined to do. Go to, let us go down and confound their language that they may not understand one another's speech. So the LORD scattered them abroad from thence ... Therefore is the name of the place called Babel; because the LORD did there confound the language of all the earth.'

Genesis 11:2–9

A simple explanation of how structures like walls, towers and so forth stand up, or perhaps more importantly, sometimes fall down, ought to be a prerequisite to any textbook on architecture. The reader who has persevered with this book so far may therefore feel tempted to ask why on earth the next chapters were not put in right at the beginning. The reason is simple. The LORD has not only confounded our language. Faced with a mention of science or mathematics, some people will shy away, accusing the writer of talking a strange tongue. They will condemn the subject as either boring or incomprehensible, and not read any further. Actually it

Roman Building Techniques

is mostly common sense; in fact a great deal of it has already been taken for granted.

A 'quality control' test that is commonly today applied to bricks is to put them into a hydraulic press and to measure what force is required to crush them. This destructive procedure is relatively easy to perform but it does not seem upon reflection to be a particularly sensible one. The compressive strength of building materials is usually unimportant. Properly employed, bricks simply can't get crushed. If it were possible to build a straight-sided tower consisting of modern common bricks, it would be 2 kilometres high before the bottom bricks were in any danger of being crushed by the weight of those above them. Built of granite blocks the same size, the tower might be at least thirty times as high!

The real difficulty with building the notional tower would be to place the blocks accurately one on top of the other, so that the structure doesn't topple. Another would be to ensure that each block was in contact with the one below it over its whole horizontal surface area. To achieve this contact is the principal function of mortar (Chapter 9). It is helpful to produce a simplified, diagrammatic and non-mathematical analysis of what happens in masonry structures. This is based on five assumptions which are found to be true in practice.

1. The material used is distorted by forces acting on it. A physicist would say it is *elastic*; although this might be misunderstood by the layman. This is true of *all* solids, even if the effect is not obvious to the naked eye.
2. The material used will not be broken by crushing.
3. Compressive forces are evenly spread in the joints, either by mortar or by accurate shaping of the blocks.
4. Friction in the joints will prevent the blocks sliding.
5. The joints are easily pulled apart; that is, they have negligible 'tensile strength'.

So even if the LORD had not confounded tongues of the men of Shinar, their tower could not have been built, at least not as a straight-up-and-down structure. The real problem would be one of stability; the structure would topple during construction. This was obvious to the medieval painters of

Strength & Stability

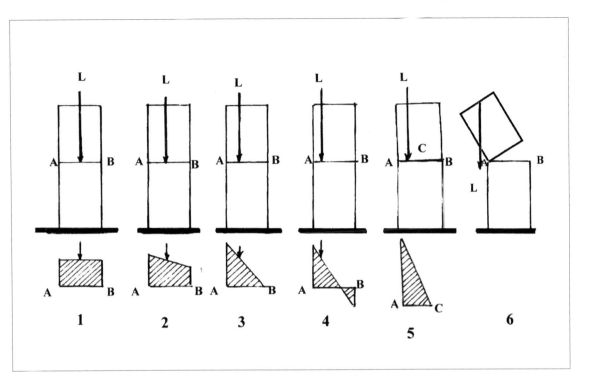

Fig. 74 Stability. Diagrams showing the simple case where a thrust line, L, passes straight down a wall or column. The distribution of the load is shown by the shaded area under each diagram: 1: with the load in the centre a compressive stress is evenly distributed across AB; 2: if the load moves to the left more of the stress is placed on the left of AB; 3: when the thrust is at the edge of 'the middle third' of the structure there is no load on the right-hand edge; 4: if the thrust is moved further to the left, the right-hand side would go into tension; 5: instead, a crack, BC, appears; 6: the top of the structure topples.

biblical scenes who showed the Tower of Babel as a variety of step pyramid or a snail-shell spiral.

This graphical analysis introduces the idea of the 'thrust line', a line passing down a structure from top to bottom that defines the position at which the vertical thrust can be considered as acting on each successive joint. For a simple, symmetrical wall, the thrust line is straight down the middle, as in Fig. 75A.

A sloping (pitched) roof, for example, poses a structural problem: its weight causes a thrust outwards, as in B. If supporting members are provided that go down to the ground, the problem is solved at the expense of floor space. In most buildings the weight of the components of the wall are successively

Roman Building Techniques

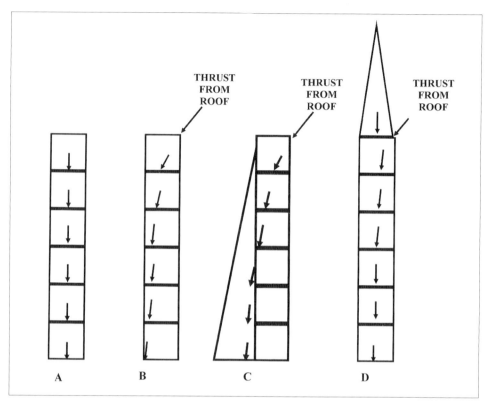

Fig. 75 Thrust lines in a wall. The weight of the structure directs (and increases) the thrust. In addition to the strategy of thickening the wall, there are two common methods of preventing its 'getting outside the middle third': either by using buttresses, or by adding weight, e.g. in the form of a pinnacle or making the wall higher above the line of the thrust.

added to the oblique thrust causing it to take a curved path as it goes down. The aim of the builder, even if he was only aware of it instinctively, was to ensure that it did not reach the outer surface of the wall, which, as seen above, would be likely to cause it to collapse.

There are several ways in which the thrust line can be made to behave safely. One is to thicken the wall, or to add buttresses, as in C. Another, perhaps not so obvious, is to add weights to the top of the wall, perhaps in the form of pinnacles or statues, as in D.

In the case of a pitched roof of moderate span, the outward thrust can be eliminated by using a tie beam. It seems surprising that Greeks, whom we always associate with geometry, triangles and trigonometry, failed to invent the tie beam. A triangular roof truss seems to us to be a fundamental necessity when constructing a pitched roof. Without it the rafters will push outwards on the walls.

Unfortunately, the roof structures of early Greek temples do not survive and, bearing in mind the unsuitability of stone for making long beams and the shortage of timber in classical Greece, it is difficult for us to guess how some of them were constructed. One theory is that on early archaic temples, roofs were actually supported by a huge heap of earth or clay piled on top of a flat wooden ceiling! We are taught to believe the ruins of classical Greek temples are aesthetically admirable when in fact they are structurally pitiful, consisting, like Stonehenge, of posts and lintels. And often the lintels (architraves) were not strong enough! Since they had no tie beams (Chapter 13), their roofs had to be supported by rows of columns carrying an architrave on which stood smaller columns parallel to the ridge (Plate 24). In several examples a row of columns actually runs right down the centre of the temple (Fig. 76). Perhaps this was the solution adopted by Solomon, with three rows of pillars? The Romans probably introduced the tie beam in the Hellenistic period. Vitruvius says (IV, iv, 2) that if the width of the temple is to be more than 40 feet, columns should be placed inside the temple and opposite the columns of the portico. He must have considered 40 feet (7.7 metres) as the maximum practical length of a tie beam, which, of course, had to be made of wood, which is relatively strong on tension (Chapter 5).

The above analysis has assumed that it is only the joints in a wall that have negligible strength. In walls, even with modern cement mortars, they are usually very weak, but, although building scientists are happy to crush materials like bricks and concrete – to find their compressive strength, they don't often try to pull them apart – to measure their tensile strength. This is only partly because it would be more difficult to do, but mostly because

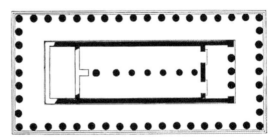

Fig. 76 Plan of the temple called 'The Basilica' at Paestum. Showing the row of columns down the middle.

they are so weak in tension that it wouldn't be worth it! The best thing to do is to avoid situations in which they are likely to become stretched, for example over openings such as doors or windows.

26

LINTELS

The harder you pull it, the longer it gets.

Schoolboy's version of Hooke's Law

Stonehenge provides a large but primitive example of a stone post-and-lintel structure. Posts are upright members of a structure. Lintels are horizontal ones that can be supported by posts. A Greek temple, despite its perceived aesthetic refinement, is not structurally any more sophisticated than Stonehenge. Lintels can be used to bridge openings such as doors and windows, and structurally they are the same as beams (Plate 25).

In the distant megalithic past it must, of course, have been found that stone lintels incorporated into masonry are prone to cracking. Four solutions to this problem could be employed. The obvious ones were to reduce the span or to use a thicker stone. The less obvious ones were to reduce the load on the lintel by somehow creating an opening in the masonry above it or by inserting something like an arch to 'spread the load' or 'carry the weight' into the masonry above the lintel.

Robert Hooke, who invented the concept of elasticity around 1676, is known to all physicists for his eponymous law. He realised that all solids must resist weights or other mechanical forces acting on them by 'pushing back' with an equal force. This became implicit in Newton's third law, published about ten years later: to every action, there is an equal and opposite reaction. Hooke saw that solids provided this by changing their shape, the change being directly proportional to the force applied. It is unlikely that the average person today appreciated the causes of the fracture any more than the builders of the gates at Mycenae: that lintels are imperceptibly bent by the load imposed upon them; that bending causes the stone to stretch and that stone is very weak in tension.

Since ancient architectural design was more a matter of aesthetics and conservatism than science, when a larger building was required, it was simply a scaled-up version of a smaller one. That is why you can't tell the size of a Greek temple in a picture without a scale (Fig. 78). If a temple was twice as big, the Greeks made the columns twice as thick, the architraves twice as long and twice as big, and so on. This took care of a major problem that they probably hadn't even thought of: the stability of the structure. If it works on a small scale, it will work on a large one (Fig. 77).

However, when classical temples became much larger than their prototype, the wooden hut, the scaling-up process failed to take into account something that was eventually articulated 2,000 years later, by Galileo, known as the 'square-cube law'. If you double the size of a block of stone, its volume, and therefore its weight, increases eightfold (Fig. 79). If the block is used in, for example, a wall, or as a column, this is usually structurally unimportant, since, as already stated, stone is very strong in compression. The architects

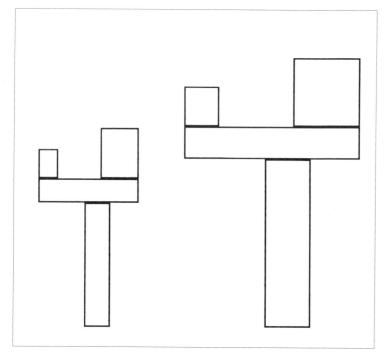

Fig. 77 Demonstration that if a structure is stable, it will remain so when enlarged.

Lintels

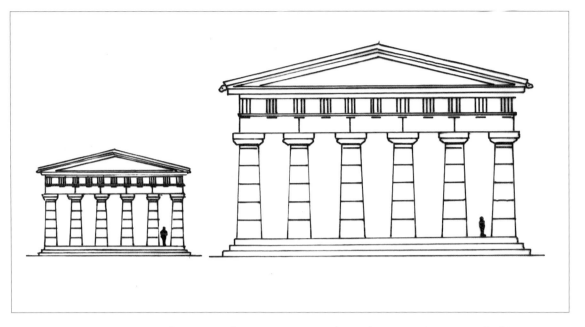

Fig. 78 Two Doric temples. Because the proportions were always the same, only an external subject, in this case a human figure, tells you how big they are.

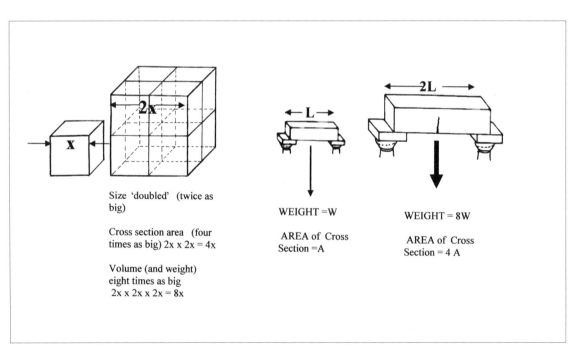

Fig. 79 The effects of the square-cube law. Doubling the size of a component increases its weight by a factor of eight, but its load-bearing capacity by only four.

Roman Building Techniques

must have realised this because, in order to reduce the work of making the joints in their monumental architecture as thin as possible, they frequently hollowed out the middles of the top and bottom faces of their blocks and column drums (*anathyrosis*), leaving just their perimeters to be dressed smooth. This meant that the weight of these, and any imposed load was, in any case, always supported on a small area.

When a block of stone is used as a lintel or an architrave, however, a bending force is imposed on it by its own weight plus the rest of the structure above it. If the size of the whole structure in which it is employed doubles, this force increases by a factor of eight, but its area of cross-section is only increased four-fold. The stress (force/area) is therefore doubled. The prototype architraves were of wood, but in tension marble has only one twentieth of the strength of wood. This is why there seems always to be at least one broken architrave in every temple (Plate 26). The Greeks, whom we always thought of as lovers of beauty, ignored the recurrent fracture of architraves, and either on aesthetic or on religious grounds were unable to change the accepted proportions of their temples. Architraves were eventually doubled (one in front of the other) but this only doubled their area of cross-section. Luckily a fracture does not necessarily cause a collapse; the broken halves become compressed, as a sort of inverted arch (Fig. 80D).

The problem of fractured architraves was increased if the supporting columns were moved further apart, and it is interesting to note that this was recognised by Vitruvius, probably in copying Etruscan proportions. The Etruscans did not build in stone.

> It is not possible to build an araeostyle temple (one where the columns stand further apart than is desirable) with epistyles [architraves] in stone or marble. Instead, wooden beams must be placed all round on top of the columns.
>
> III, iii, 5

One of the ways of bridging a space, or of reducing the load on a lintel, is to use corbelling; where successive courses of stone, held in place by their neighbours, are jutted into the space. The general effect, depending on the size and shape of the stones, is to produce a triangular opening (Fig. 80A). A similar opening can be produced by a pair of stones resting together (Fig. 80C).

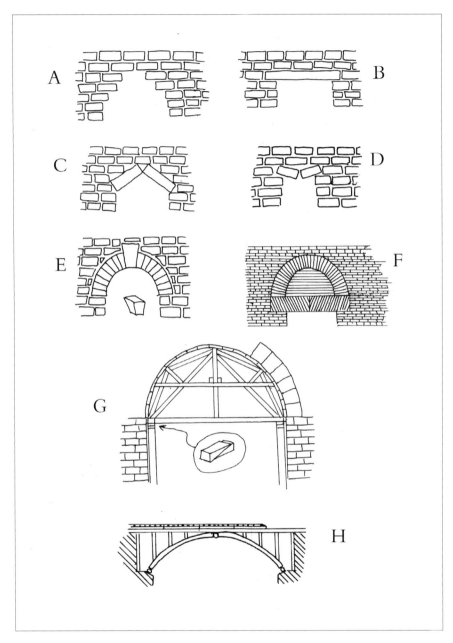

Fig. 80 Gaps in masonry: A: corbelled; B: lintel or architrave; C: chevron lintel; D: this arrangement sometimes occurs by accident in brickwork. The bricks over the gap are 'prestressed'. It is similar to a broken lintel. E: a voussoir arch. F: lintel with 'relieving arch' to direct the load to the ends. In Roman 'brickwork' this arrangement is often superficial and when facing bricks fall off, solid concrete is exposed behind the arch. G: construction of a voussoir arch. The individual voussoirs are laid onto the 'centring' or 'formwork', the height of which is often adjusted by pairs of wedges (*inset*). H: a modern three-hinged arch. A broken lintel or architrave is functionally similar.

A common expression used in modern engineering is 'reinforced'. Iron rods or bars (in modern jargon 'rebars') are cast into concrete to strengthen them where bending forces are likely. On lintels and beams they have to be near the bottom, where stretching would occur. The use of reinforced concrete dates from the middle of the nineteenth century. Although the Romans used iron clamps to secure stone blocks in place, they did not have the idea of putting iron in concrete to reinforce it. In any case, their concrete was not as dense and impermeable as modern material, which uses Portland cement. Their impure iron was susceptible to corrosion and was cased in lead.

Engineers make use of the term 'prestressed'. Prestressing is the process of imparting to a structure a compressive stress in places where under a working load it would experience destructive tensile stress. This is experienced in the simple process of taking a row of books from a bookshelf and holding them with a hand at each end of the row. The row itself has no tensile strength and, if not supported by the shelf, it would fall apart. By applying pressure by a hand at each end the books can be made to act as a beam. The pressure of the hand sets up a compressive stress which overcomes the tensile force that would cause the books to fall apart. Prestressing can be seen, produced deliberately or accidentally in many places, such as in partially collapsed brickwork, or where a lintel is broken (Fig. 80D).

Modern pre-stressed concrete is made by tensioning rods or wires, which pass through it, effectively putting the material in compression. This technique has only been used since the middle of the twentieth century and was therefore not available to the Romans.

27

ARCHES & VAULTS

> I will set my bow in the heavens.
>
> Common misquotation of Genesis 9:13

We derive the word 'arch' for a curved symmetrical structure used to span an opening from the Latin *arcus*, a bow. A vault (from *vulvo*, 'I jump') is an arch extended along its horizontal axis. The true arch is the epitome of the use of prestressed stone. Each of the components, called a *voussoir*, is wedge-shaped, and fits against its neighbours to create part of a circle (Figs 80E & 80G).

An arch has to be constructed on a temporary formwork or centring, usually made of wood. The centring can be supported from the ground, but this requires extra timber, and obstructs the opening during work. In some cases, therefore, especially where the arch or vault is a long way from the ground, as with aqueducts, a projection was provided in the abutment or pier to support it. This became incorporated as part of the visual presentation as impost moulding, the capital of a pilaster or, in ceiling vaults constructed of stucco or plaster, on laths, reeds or the like, as part of a cornice. Support for the centring was often provided in bridges by making the piers project inside the arches. The temporary wooden structure was removed once the final, top *voussoir* was in place. Although its structural function is exactly the same as that of the other *voussoirs*, this one was customarily slightly larger then and is often called the keystone. Removal of the centring allows the *voussoirs* to drop together and the structure is put into compression by their weight. If the *voussoirs* are properly shaped, no mortar was needed. As building progresses, more weight is put onto the arch, but, as already pointed out, stone is strong

Roman Building Techniques

Fig. 81 Mud brick vaults of the store-rooms of the Ramesseum, on the West Bank, Luxor, Egyptian, thirteenth century BC.

in compression. Because the shape of the *voussoir* is a wedge, the true arch exerts a sideways thrust and it needs lateral restraint by abutments of masonry or earth on either side (Plate 27).

The Egyptians used true arches in tombs as early as the sixth century BC, for example at Maydum, but did not employ them in monumental, visible, architecture. Barrel vaults can be seen in the wide storage chambers of mud 'brick' at the Ramesseum on the West Bank at Luxor (Fig. 81) about 1250 BC. At that time, however, builders seem to have made stone vaults by corbelling rectangular blocks and then shaping the intrados – the inside of the vault – afterwards. Voussoir arches and vaults were used in the Hellenistic world in the third century BC for structural purposes and it is suggested that the Romans inherited their use as architectural features. Early arches were semi-cylindrical, because flatter arches, based on arcs of cylinders, cause greater sideways thrust, but they were eventually used as the builders gained confidence in the design of suitable abutments.

The introduction of concrete, which, placed on suitable wooden centring or formwork, allowed arches and vaults to be built without the laborious production of shaped blocks, permitted the construction of, for example, Porticus Aemila, a warehouse in Rome that was built about 196 BC consisting

Arches & Vaults

Fig. 82 The arch of Septimius Severus, Rome. Although it has the appearance of three true voussoir arches, with keystones, the absence of buttressing makes it clear that this, like other triumphal arches, is, in fact, a block of concrete pierced by openings.

of six barrel vaults side by side, connected by arches to cover a continuous space 487 metres by 50 metres. Concrete arches and vaults effectively became monolithic structures and exert little sideways thrust. The so-called triumphal arches are effectively large blocks of concrete with holes through them! They are made to look as though they are ashlar masonry, with *voussoirs* and keystones, by means of a veneer facing of stone (Fig. 82).

Where it was required that two masonry barrels or tunnel vaults of the same height should intersect at right angles the shaping the *voussoir* blocks at the groin (the sharp junction of the curved surfaces) would have posed very complex three-dimensional geometric problems that the Roman masons preferred to avoid. Their solution was to make the invert of one vault lower than the other, so that it opened entirely in its vertical wall (Fig. 83). Groined cross vaults resulted when barrel vaults intersected at right angles at the same

Roman Building Techniques

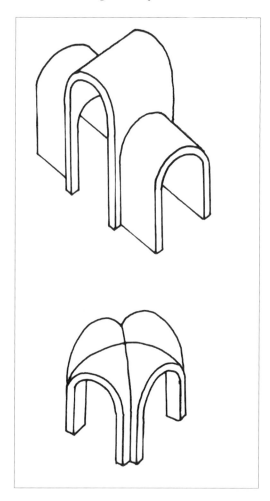

Fig. 83 Top: Two vaults crossing one another. If they were made in stone it would be difficult to shape the 'groins'. The solution adopted was for one vault to cut the other in the vertical walls. Bottom: Once concrete construction was adopted, the problem was passed to the carpenters who made the formwork, and the vaults could cross at the same level.

height. Here the thrust lines (Chapter 25) are directed to the corners where they are taken into piers so that the walls are no longer necessary to provide abutments.

Once the carpenters had undertaken the task of creating the three-dimensional space, removing it from the masons, it was discovered that many shapes of vaults or domes could be constructed comparatively simply to cover rooms of different plans (Fig. 85).

Greek theatres and stadia were built into hillsides, which provided the support for the tiers of seats. The use of shuttered concrete made it possible to construct new versions of these and the typically Roman amphitheatre from the ground up. The Flavian amphitheatre in Rome, whatever the true

Arches & Vaults

origin of its common name, 'the Colosseum', is a truly colossal structure. Most accounts of it express astonishment at its size, and at the short time it took to build, and wonder at the organisation that must have gone into this. What is usually forgotten is the achievement of planning its geometry, and of communicating it to the builders (Fig. 84).

The Romans often incorporated ribs and arches of building brick into vaults and domes. What function their builders intended them to perform, and what, if anything, they actually did, is not at all clear. Unlike the ribs in later medieval vaulting, they did not support the panels between them. Ideas

Fig. 84 A section through one side of the Colosseum.

145

put forward have included stiffening the structure and, more convincing, dividing it into discrete areas, each of which could be filled separately with concrete. However, many of the apparent 'relieving arches' over lintels in concrete walls are merely superficial, cosmetic, surface features (Figs 80D & Plate 9).

28

DOMES

My intentions were that this sanctuary of All Gods should reproduce the likeness of the terrestrial globe and of the stellar sphere.

Frequently attributed to Hadrian, about the Pantheon

There is a Latin word for a dome, *convexum*, but we use a word derived from the Latin for a house, *domus*. It was first used in English by the gentry in a tongue-in-cheek manner in the sixteenth century to describe their large houses. Sir Thomas Wilson, author of *The Arte of Rhetorique* wrote a letter in 1553, 'Dated at my dome, or rather Mansion, in Lincolnshire'. In other languages, it became attached to religious buildings, regardless of their shape, as *duomo* in Italian, or *dom* in German, for a cathedral, implying the House of God.

It seems that almost any curved solid covering for a structure can be called a dome, although it might be preferable, perhaps, to use 'vault' for structures that have simple linear symmetry, and 'dome' for those with rotational symmetry. Some common types of Roman vaults or domes are illustrated in Fig. 85.

Geometrically, a hemispherical dome can be seen as an arch, rotated about its vertical axis. It might, with skilled shaping by the masons, be constructed using tapered blocks analogous to the *voussoirs* of an arch, but the time-consuming and difficult work was not essential. A hemispherical dome can be built, like an igloo, by laying a series of horizontal rings of material each of which is, in effect, corbelled from the one beneath. It does not then require centring or formwork, as each ring or zone acts as a sort of horizontal arch and also puts the ones below it in increased compression. Unlike an arch, the top of a dome can therefore be left open as an oculus or eye to admit light.

Roman Building Techniques

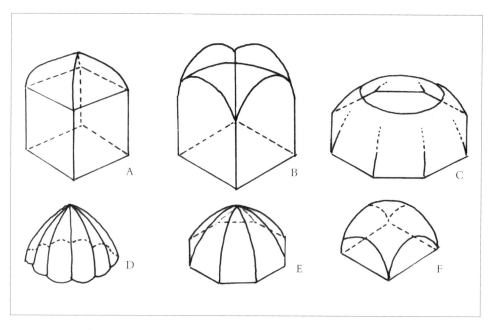

Fig. 85 Types of vaulting: A: pavilion or cloister; B: cross; C: octagonal domical merging to a spherical dome (Nero's Golden House); D: umbrella or parachute; E: octagonal Domical; F: sail.

Early Roman domes were constructed on cylindrical, drum-shaped buildings, the walls of which were either buttressed or were thick enough to contain the imposed thrust. The Pantheon, built by Hadrian, is the most impressive and arguably the most beautiful interior ever conceived (Plate 2). The coffered dome is 43.7 metres in diameter; its unglazed oculus 8.3 metres across. It shows an empirical appreciation and anticipation of the problems that might be imposed by such a structure, in particular by the choice of lightweight aggregate where strength was less important, and the thickening of the dome at its equator (Fig. 86). The drum was made structurally thick enough to take the outward thrust of the dome while saving material by means of the doorway, four square-sided bays and three apsidal ones. In the thickness of the piers between the bays there are voids in the form of half-domed chambers.

The use of a cylindrical drum worked well for free-standing structures, but difficulties appeared where the dome was, for example, covering a room that was an integral part of a larger building, since this inevitably created small, odd-shaped spaces with sharp internal corners. This difficulty was to some

Domes

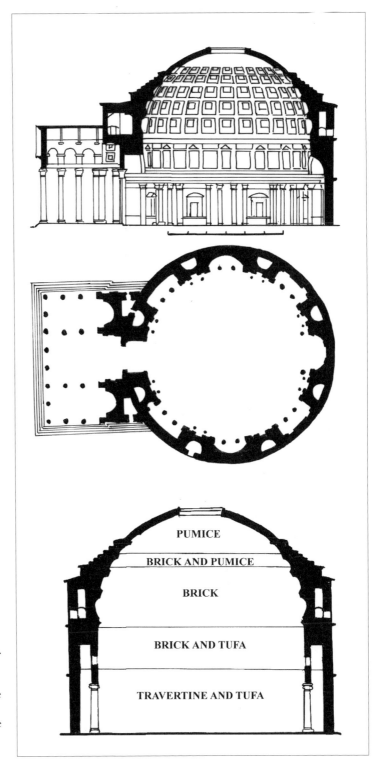

Fig. 86 Sections and plan of the Pantheon. The bottom section shows, diagrammatically, the principal materials used in the successive layers of the structure.

Roman Building Techniques

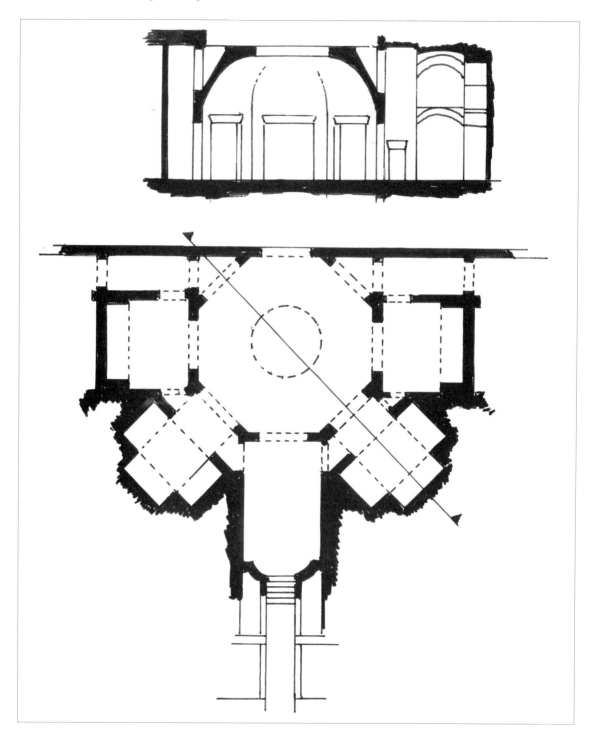

Fig. 87 Section and plan of the octagonal room of Nero's Golden House, AD 64–68. Light wells over the dome allow light into the surrounding rooms.

Domes

Fig. 88 Building blocks for byzantine architecture. Centre shows the geometry of squinches: A: dome, B: drum, C: square tapered to an octagon, D: square modified to octagon by squinches, E: square to circle by pendentives, F: dome to fit E. F–H: squinch, drum and dome; I–J: vaulted nave; K–L: vaulted apse; M: polygonal core for example San Vitale, Temple of Minerva Medica.

extent avoided by putting the domical top on a polygonal, frequently, an octagonal plan, as in Nero's Golden House, which is a wonderful example of ingenious use of early brick-faced, concrete architecture. The 'domical vault' of the octagonal room springs from eight piers that were originally decorated with stucco and marble pilasters as a facetted umbrella, which changes upwards into a spherical dome with a wide oculus to admit daylight. The rooms awkwardly packed around the octagon are lit by openings over the top of the vault. (Fig. 87).

Various methods were employed for putting a spherical dome on a square plan. One was to make the transition of the square into a thick-walled octagon upon which the hemisphere would fit. A bridge could be made across the corners of the square. Another version of this was the squinch, which was created by filling the bridge with a hemi-dome (Fig. 88D).

151

Roman Building Techniques

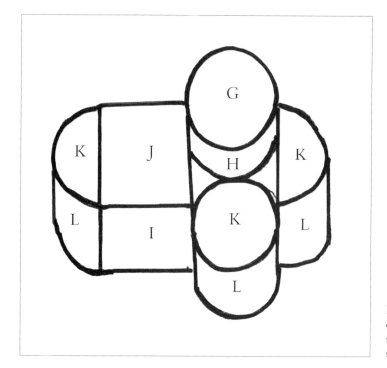

Fig. 89 Byzantine church made by a student from the units shown in Fig. 88.

The elegant solution to the problem was the development of the pendentive, which is a surprisingly simple concept. It is, in effect, the lower parts of a dome with a diameter equal to the diagonal of the square in which it sits. Any horizontal section of this incipient dome where its diameter is equal to, or less than, the length of the side of the square will be a circle, on which a dome or a drum can be constructed (Fig. 88E).

In chapter 26 it was shown how the use of concrete made it simpler to construct cross-vaults, and how this, in turn, liberated space, because vertical walls could be eliminated. A similar liberation of space became possible when pendentives were introduced, since their lower margins were arches, and walls were not needed under them. In this case, however, the outward thrust of the dome must be countered by buttresses or, for example and more excitingly, by adding a half dome. Byzantine architects exploited the possibilities by designing buildings which drew together several architectural elements, and possible outcomes can be explored using the solid block models illustrated in Fig. 88, which can easily be made on a lathe. Fig. 89 illustrates one possible outcome.

152

Domes

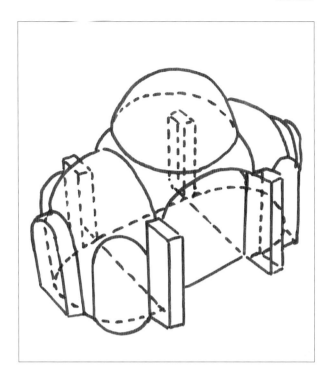

Fig. 90 Structural analysis of Hagia Sophia, completed AD 360.

A visitor to Haghia Sophia in Istanbul cannot fail to be amazed. The architects, Isadore of Miletus and Anthemius of Tralles (said to be 'physicist' and 'mathematician' respectively), have created a space roofed by domes, half domes and vaults, which are invisibly supported. It is like the inside of a hot-air balloon, about to take off and float into space (Fig. 90).

Drums did not need to be solid, and lightening of the drum of the Pantheon has been described above. In the caldarium or heliocaminus of the Baths of Caracalla (Fig. 73) we see large openings in a structure supporting a dome. The outward thrust required thickening of the drum into heavy piers between the openings. In Rome, the so-called 'Temple of Minerva Medica', probably a nymphaeum, built at the beginning of the fourth century, had openings in an octagonal drum which were buttressed by a series of apsidal niches (although it later required external support, Fig. 91). The weight of the dome of the Mausoleum of Constantine's daughter, Santa Constanza, in Rome is largely taken by an arcade, and the outward thrust by an annular barrel vault. In San Vitale, Ravenna, we see the development of this idea. There the weight of the dome was greatly reduced by constructing it with earthenware pots.

153

Roman Building Techniques

Fig. 91 Plans and sections of concentric cored buildings at the same scale. Left to right: the Temple of Minerva Medica (a late fourth-century nymphaeum), Rome; Santa Constanza, Rome, AD 330; San Vitale, Ravenna, AD 526–47.

Fig. 92 Syracuse Cathedral.

Domes

It was protected by a timber roof, as were many medieval structures. This reminds us that the practice of dividing the contents of books on architecture into 'periods' might lead to the idea that 'Roman' architecture ended, and 'Byzantine' began, whereas it is obvious that the builders, the materials and the laws of physics did not change arbitrarily at a certain date.

The basilica, which was derived from the revered Greek tradition (Chapter 2), had always functioned as a large interior thought to suit the dignity of authority, and it became the accepted design for Christian churches in the Western Empire. For this reason, the Western liturgy became theatral, with the congregation in the position of an audience, and the priest in that of the actor. In Syracuse Cathedral you can see a not-too-successful attempt to turn a pagan temple inside-out, to create a rather cramped basilica. To form aisles the spaces between the columns were blocked, and arches were cut through the walls of the *naos* (Fig. 92).

In contrast, the architectural tradition in the Eastern Empire became based on the dome. However, the church of Hagia (Santa) Irene in Istanbul, close to Hagia Sophia, begun after AD 532 to replace an original basilican church, illustrates the beginning of the development, in the Middle Ages, of the domed basilica (Fig. 93).

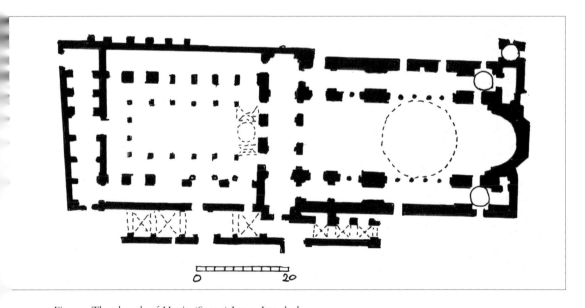

Fig. 93 The church of Hagia (Santa) Irene, Istanbul.

ACKNOWLEDGEMENTS

I began accumulating the ideas and illustrations in this book over fifty years ago. It is therefore impossible to remember to thank all the large number of people who have contributed to it and to whom I am very grateful.

Apart from formal training in science and mathematics, most of my knowledge has been picked up 'on the job', either as a building technologist for the Central Laboratories of George Wimpey & Co. and the Chalk Lime and Allied Industries Research Association (CLAIRA) or (bless them for their confidence in me!) as a tutor for the Extramural Departments of London, Cambridge and Essex Universities, where I learnt Greek, Roman and Egyptian archaeology *with* the students. I did research on Roman baths at the Institute of Archaeology, University College, London, supervised by Richard Reece, where Professor K. D. White and Mark Hassall discussed practical problems with me.

My special thanks must go to my colleagues at Wimpey's and at CLAIRA, especially the labourers and the craftsmen who did the real wok and taught me their expertise. Practical support in my Roman studies was provided through the good offices of J. B. Ward-Perkins and Professor Musso. Fieldwork was assisted by Frank Sear, Bryan Scott, my wife Merle and my daughter Sylvia. I would like to thank everyone at Archäologischer Park Xanten, especially H. J. Schalles.

The illustrated manuscript was read critically by my wife, and by Nick Tracken and by Mark Hassall, each of whom, especially Mark, suggested improvements and additions. I am most grateful to Hazel Bell, who not only provided the index, but also made many helpful suggestions. Any surviving errors and omissions must, of course, be mine.

Unless otherwise stated, all the photographs and line drawings are my own, although many of the latter were derived from authoritative sources. I hope their authors will forgive me. I have used and paraphrased several translations of classical texts; I trust I have not infringed copyright.

FURTHER READING

Information has been provided by a number of works which are out of date or out of print, with titles like *Building Construction*, *Building Educator* and *Practical Building*, and inspiration from, among others, Bannister Fletcher's *A History of Architecture* and R. E. M. Wheeler's *Roman Art and Architecture*. J. E. Gordon's *Structures, or Why Things Don't Fall Down* inspired and provided much of what I have written about the realities of building, and J. G. Landel's *Engineering in the Ancient World* is a useful introduction to the mechanics of the Roman Period.

The most important modern book on Roman building technology is *La Construction Romaine* by Jean-Pierre Adam, which has been translated into several languages, including English. My more light-weight volume is intended to provide an original way of looking at the subject.

INDEX

Note: Page references in italics indicate illustrations and diagrams. Numbers in bold type indicate plates, in the section between pages 96 and 97.

Adam, Jean-Pierre: *La Construction Romaine* 102
adobe walls 46
agora 15, 18
Agricola 14
Agrippa 107, 127
Alberti, Leon Battista: *De re Aedificatoria* 14
amphitheatres 21, 22, 23, 144–5
Anthemius of Tralles 153
antefixes 71, 72
Antonine Wall 49
Apollinaris, Sidonius 105
Aqua Sulis 108
aqueducts 28, 54, 99–101, *101, 102*
  Anio Verus 99
  Appia 99
  Pont du Gard **16**
arches *139*, 141–6, **27**
  honorific 24
  of Septimius Severus *143*, **4**
  triumphal 24–5, 143
  see also bricks/*voussoirs*;
  bridges; domes
Archimedean screw pump 96, 97
architraves *see* lintels
Aristophanes: *The Birds* 31
ashlar masonry 54
Athens 21
Ausonius: *Mosella* 40, 114

basilica 18, *19*, 71, 155
  Fano 13, 18
  Lateran 89
  plan *133*
baths 10, 24
  activity in 111, 126–8
  buildings in Britain 8–9, 109–12
  caldarium (hot room) 24, 109–11, *115, 117*, 125, 127, *128*, 153, **20**

chimneys 114–21, 122
  heating/hypocaust 109–13, 122–5
  at Pompeii 24, 112, 114, *115, 116, 117*, 118, 119, *124*, **19, 20,** 21
  public (*thermae*) 109–13, 121, 125, 126–8
  Thermae Antoninianae 127–8
  Villa Silin 121, **22**
Baths of Caracalla 125, 127, *128*, 153
Baths of Diocletian 127, **23**
beams 15–16, *17*
  tie beam *17*, 18, 70, 132
Bearsden fort baths 119
Bible 15, 36, 91, 129
blacksmiths 92–3, *93*
bobbins 119–20, *119*
Boscoreale, villa
  baths *110*, 124
  door 90, **15**
braziers 111–12, 122, **19**
bricks 47, 54, 61–3
  mud 47, 64
  sizes 63
  strength 129–34
  thickness 64–5
bridges 27, 35, 44
  Pons Aelius **5**
  Pons Fabricius **6**
Britain 33, 36, 46
  bath sites 8–9, 109–12
  housing 47–9, 70, 82
Byzantine architecture 66, *151, 152*, 152–5

Caesar, Julius 13, 27, 35
Casa del Bracciale D'oro **8**
Casa dei Pittori al Lavoro 58
Castel Saint'Angelo, Rome **5**
Cato: *De Agri Cultura* 50–1
CBM 63

ceilings 57, 133
cement 54, 55
charcoal 35, 36, 122
Chersiphron 45
Chesters 121
chimneys (flues) 114–21, *117*, 122-4
chorobates 32
Church of Sta Maria degli Angeli 127
churches, Christian 18, *151, 154, 155*, **23, 28**
circus 23
  of Maxentius 22
cisterns *103*
clay 46, 47, 55, 61
cob 46
Colosseum (Flavian amphitheatre) 22, 144–5, *145*
colours 59–60, 79–80, *81*, 84
columns 8, 15, 16, 19, 24–5, 42, *43*, 45, 58, 63, 136, *137, 138, 154, 155*
  foundations 34
  thrust line *131*, 133, *133*, 136
  Trajan's *frontispiece*, 25, 27
concrete 21, 23, 38–9, 54, 65, 66, 142–3, 144
  floors 77–8, 82
  reinforced 140
copper 39, 60, 92, 94, 108
coppicing 36
corbelling 138–9, *139*, 142, 147
Cornwall 94
cranes 42–5, *44, 45*
cranks 40, 41, 42
Ctesibius 40, 98, 104
Cyprus 94

daub 47–9, **7**
dioptra 31–2
domes 147–55, *148, 149, 150, 151, 151*
  pendentives *150, 152*

domestic buildings 29
Domus Tiberiana **12**
doors 89–90, *90*, **14**, **15**
Doric architecture 15–16, *17*,
  137
dyes 60

earthworks 25–6, 49
Egypt 47, 142
Etruscan architecture 16, 138

Fano basilica 13, 18
Flaccus, Siculus 27
floors 73, 77–81, 82, 123
  heating/of hypocaust 112–
  13, 121
  opus sectile 77–8, **12**
flues *see* chimneys
fortresses 26
forum 14–15, 18
foundations 33–5
fountains 81, *81*, 101, 104,
  18
  labri **20**
frescos 58
Frontinus 31, 105, 106, 107
Fronto, Cornelius 123
furnaces 84, 86, 109, 112,
  114, 121, *124*, 125

Galileo 136
granaries 77
glass 84–8, 91
  blowing 85–6, *87*
  frit 60
  slag 27, 91–2
  tesserae 78–81
grills 84, 85, *124*, 125
groma 31, *31*
grotto 81
gymnasium 24
gypsum plaster 52, 58

Hadrian 147, 148
Hadrian's mausoleum 25, **5**
Hadrian's villa, Tivoli 121
Hadrian's Wall 44, 49
Hagia Irene church, Istanbul
  155, *155*
Hagia Sophia, Istanbul 153,
  153
Herculaneum 10, 39, 77, 81,
  83, 85, 89
  baths 114, *115*, *118*, 119,
  125
Herodotus 111
Heron 30
Hierapolis 40
Hittites 92
hodometer 30
Homer: *Odyssey* 93
Hooke, Robert 135
House of Diana, Ostia 66, **9**
House of Julia Felix, Pompeii
  117

House of the Labyrinth,
  Pompeii 119
House of the Large Fountain,
  Pompeii 81
huts 47, 48, 73–4, 136
hypocaust 24, 36, 112–13
  chimneys 114–21, 122–4

impluvium 95
Institute for Archaeologists 7
iron 36, 61, 92–4, 140
  oxide 60
Iron Age 48, 70
Isadore of Miletus 153

Jericho 47
joints
  brickwork 130, 133, 138
  pipes 100, 102, *106*, 108, *110*, 138
  woodworking *38*, *39*
Juvenal: *Satires* 66

kilns 50–2, *51*, 61, *62*, 73

ladders 82–3
lateres *see* tiles
laths 57
latrines 28, *28*
lead 39, 94
  pipes 100, 105–8
Leptis Magna, Libya 40
  Arch of Septimius Severus *143*, **4**
  baths *120*, 120–1
  Macellum **3**
  roofs 76
  Villa Silin 121, **22**
light and sun 88, 122, 125,
  127, 151, **20**, **28**
lime 50–2, 56, 60
  mortar 50, 53, 54–5, *68*
  putty 52, 56–7
limestone 50, 54, 84, 91
lintels (architraves) 63, 89,
  133, 135–40, *139*, **25**
  post-and-lintel 8, *133*, 135
Little Woodbury 48
locks 89, *90*
Londinium 40
Lullingstone villa 46, 119

macellum 19, **3**
marble 15, 19, 23, 40, 45, 60,
  69, 81, 84, 138, 151
  dust 56, 78
markets 14–15, 19, **20**
  Trajan's **20**
Martial: *Epigrams* 87–8
Mausoleum of Santa
  Constanza, Rome 153, *154*
Maydum 142
measurement 30–2
Mencken, H. I.: *The History of
  the Bathtub* 112
Metagenes 45
metals 60, 91–4

Meton 31, 32
military works 25–6, 49
mines 97
Mitchell, Charles E.: *Building
  Construction* 67
Mithras 18
mortar 8, 53–5, 130
  lime 50, 53, 54–5, *68*
  in roofs 71, 72
mosaic 58, 77, 79–81, *81*, **28**
mud 46–9
  bricks 47, 64
  see also clay
Mycenae: gates 135, **25**

Necropolis 93
Nero's Golden House 148, 150,
  151

oak 38, **7**
Olympia 15, **27**
Oplontis 57
opus caementicium 54
opus incertum 64
opus reticulatum 65
opus sectile 77–9
opus signinum *see* concrete
opus spicatum 79
opus testaceum 63, 65, 66, **9**
Orata, Sergius 112
organ, wind-driven 40, 41
Ostia 10, 26, 125

Paestum 71, *133*, **26**
paint 58–60
Palladio, Andrea 127
Panini, Giovanni **2**
Pantheon 16, 147, 148, *149*, 153, **2**
pilae 112-14, *113*
pile-driving 35
pipes 94, 99–102, 105–8, *106*
  chimneys 114, 117
  joints 100, 102, *106*, 108, *110*
  sizes 105–7
Piranesi, Giambattista 127
plaster 39, 56–60
  gypsum 52, 58
Pliny: *Natural History* 30, 46,
  52, 60, 75, 92, 95
Pompeii 10, 15, 52, 64, 77, 83
Porticus Aemila, Rome 142–3
post-and-lintel 8, 133, 135
pozzolanic materials 54
prestressing *139*, 140, 141
pumps 40, 96–9, *98*
putty, lime 52, 56–7

quarries and quarrying 43, 45
quenching 93
quicklime 51–2

Ramesseum, Luxor 142, *142*
reeds 57
reinforcement 140
rendering 56, 57, 118

roads 27
Roman Empire 10
Rome 45
  houses 64–6, 75
roofs 70–6, 71, 72, 121, 131, 11
  gutters 95
  thrust line 131–3, 132
  see also tiles/roofing
round-houses 47, 70

San Vitale, Ravenna 151, 153, 154
sandals 111, 123, 123
sarcophagus of Aurelius
  Ammianus 40
saws 37, 38
  water-powered 40, 41
scaffolding 8, 67, 68–9
scaling-up 136–7
Scipio 122, 125
Seneca: *Epistulae Morales* 64, 122–3, 125, 126
Servian Wall 26
sewers 28
shingles 75, 76
shops 19, 20, 43
shuttering 64–6
shutters 89, 90
Sicculus, Diodorus 94
Sicily 79
slag 27, 91–2
Solomon 15, 133
Spain 57, 60, 94, 97
square-cube law 136, 137
squinch 151, 151
St Mark's Venice 28
stadia 23, 45, 128, 144, 27
stairs 82–3, 83, 13
stoa 18
stone 38-9, 40–5, 135–6, 138, 141–2
Stonehenge 133, 135
Strabo 18
stucco 57–8
stylobates 33
sun room 125
sunbathing 127
surveying 30–2
sweat baths 111
Syracuse Cathedral 154, 155

tabernae (shops) 19, 20, 43
Tacitus 14
tegulae see tiles, roofing
Temple of Aphaia **24**
Temple of Diana 45
Temple of Divus Romulus 89, 14
Temple of Hadrian 34
Temple of Hera 15–16
Temple of Minerva Medica 151, 153, 154
Temple of Vesta 16, 1
temples 15–18, 135, 136
  circular 16
  Doric 15–16, 17, 137

Etruscan 16, 138
Greek 16
Paestum 133, **26**
Roman 15–17, 1
Romano-Celtic 16, 17
  in Rome 16, 34, 89, 151, 153, 154, 14
Terracina 36–7
terracotta 61–2, 63, 66
  pipes 105, 114, 117
  see also tiles, roofing
tesserae 78–81
testudo 124, 125
thatching 75–6
theatre
  amphitheatres 22, 23, 144–5
  Greek 21, 144
  Roman 21
Theatre of Herodes Atticus 21
thermae (public baths) 109–13, 121, 125, 126–8
thrust line 131, 131–2, 132, 144
tiles 8, 61–3
  flooring 78–81
  flue (*tubuli*) 8, 9, 114, 116, 118, 118–20, 120, 122–3
  lacing courses 66
  patterns 118–19
  roofing (*tegulae*) 71–3, 74, 76, 119
  tegula hamata 119
  tegulae mammatae 114–17, 115, 116, 120
  see also bricks
timber 15–16, 17, 36–9
  charring 35
  in roofs 70–1
tin 94
tombs 25, 79
  Egyptian 142
  Etruscan 25
  of the Haterii 45
  of Trebius Justus 68
tools 93–4
  blacksmiths' 92, 93
  masons' 40, 41
  plasterers' 59
  woodworkers' 37
towers 129–31
  water towers 102–3, 103, 17
Trajan 20
Trajan's column *frontispiece*, 25, 27
translation 13
tree-felling 37
tubuli see tiles/flue
turves 49

Valerius Maximus 112
vaults 57, 120–1, 121, 141–6, 142, 148, 151
  see also domes

Vegetius 49
Vesuvius, Mt 10, 54, 90, 118
Villa Capo di Bove 78
Villa Farnesina, Rome 59
Villa of the Quintillii 10
villas 46
  Romano-British 82
  Scipio's 122
Vindolanda 123
Vitruvius Pollo, Marcus 10, 11–13, 12
  De Architectura 12, 70
    cited or quoted 11, 15–16, 17, 23, 27, 30, 31, 32, 33, 35, 42, 45, 49, 52, 53, 54, 56–7, 60, 62, 64–6, 71, 77, 97, 98, 100–1, 105–6, 111, 112, 115–16, 133, 138
volcanic ash 54
voussoirs 63, 121, 139, 141–2, 143, 143, 147

walls 53
  cavity 115, 117, 120, 122
  concrete 65
  facing 63, 64, 66, 139, 143
  and foundations 33–5
  masonry 26, 64–9
  mud 46–9
  rubble 66
  strength 129–34
  surrounding towns 26
water 28, 95–104, 105
  boiler 94, 109, 110, 124
  drainage 95
  drinking 95, 107–8
  lifting 96, 97–9
  supply 95, 100–2, 101, 102, 103, 104, 105–8, 17, 18
  see also aqueducts; baths
water level 32
watermill 40, 41
wattle-and-daub 47–9, 57
welding 93
wells 95
Welwyn, Herts 9, 82
Welwyn Archaeological Society 109
Welwyn Roman Baths 9, 109–11, 110
Wheeler, Sir Mortimer: *Roman Art and Architecture* 24, 46
wheels 30
Wilson, Sir Thomas 147
windows 84–8, 120, 122, 124, 125
wood 36–9, 57
  see also timber
woodworkers' tools 37
woodworking joints 38, 39